	Multi-sample		Two Factor
Samples	Independent Groups	Related Samples	
Correlation			
Contingency or Phi Coefficients A. p. 207 B. p. 208 Table F, p. 288	χ^2 Multi-sample A. p. 29 B. p. 42 Table F, p. 288	* (Cochran Q Test)	
Rank-Order Coefficient A. p. 79 . p. 88 Table M, p. 300	Kruskal-Wallis A. p. 67 B. p. 74 Table K, p. 296	Friedman A. p. 71 B. p. 76 Table L, p. 299	
Pearson Coefficient . p. 109 . p. 126 Table N, p. 301	Single-Factor Analysis of Variance A. p. 156 B. p. 168 Table E, p. 282	Treatments-By-Subjects Analysis of Variance A. p. 181 B. p. 192 Table E, p. 282	Factorial (Two-Way) Analysis of Variance A. p. 173 B. p. 188 Table E, p. 282

* These Tests Are Not Covered in the Text But May Be Found in Siegel (1956)

introduction
to
statistics

introduction
to
statistics

HERBERT FRIEDMAN College of William and Mary

Random House NEW YORK

ISBN: 0-394-31337-2

Library of Congress Catalog Card Number: 79–168498

Manufactured in the United States of America. Printed and bound by The Book Press, Inc.,
Brattleboro, Vt.

First Edition

9 8 7 6 5 4

Design by J. M. Wall

Cover design by David Burke

To Cele and Alfred Friedman
to whom full credit is due for
the initial conception of this project

This book is based on a one-semester course in introductory statistics that I have been teaching and modifying for most of the past decade. The emphasis of the course and the text is not on the statistical procedures per se but rather on the broader problem of analyzing and interpreting the data at hand.

The content and organization of this book are atypical in two respects: The first is the emphasis on understanding and accounting for the data rather than a concentration on the calculation of probability values; the second is that the text starts with a discussion of statistical inference and data interpretation. Descriptive statistics are taken up when they are appropriate to the discussion at hand rather than as separate topics. The basic concepts of the text are reviewed repeatedly in discussions of the appropriate measure for different situations. Students generally require considerable time to "get the knack" of hypothesis testing and data interpretation. Introducing these topics early provides ample opportunity for the student to review and assimilate them and establishes a clear framework on which to build the later chapters. In contrast, the practice of introducing inferential statistics late in the course tends to overwhelm the student, so that his comprehension and retention both suffer.

A very efficient (but sometimes disconcerting) feature of the approach in this text is that the material in each chapter does not have to be totally mastered before proceeding, since discussion of the major topics is generally carried across several chapters. For example, many key concepts, such as hypothesis testing, probability, and correlation, are introduced in early chapters; later in the text these topics reappear to be dealt with more thoroughly, but, by then, they are somewhat familiar. The student must take some details on faith until they can be explained at a future time, but this is readily accepted when the rationale is understood.

The building up of topics and the interrelationships of the chapters differ from the usual approach of assigning each topic or test to a separate and distinct chapter. The present arrangement, therefore, requires that the chapters in the text be considered in sequence. To cover a large range of topics within one semester (or trimester), the class must proceed in what sometimes may seem like a "head-long" rush, trusting to the built-in repetition and review to clarify and consolidate the material. By the end of the course most students are slightly surprised to find they understand the material and that it is reasonably "easy" to take the basic steps in interpreting data. The process resembles learning to ride a bicycle, which at first is a confusing and seemingly impossible task, but after practice, it is seen to be simple and straightforward (though—of course—not foolproof).

The text is primarily concerned with what will be retained after the course is completed, and busy work and extraneous details are held to a minimum.

The explanation and discussion of the various measures are mainly on a verbal rather than a mathematical level, designed to give the student an intuitive grasp of the subject and requiring, at most, some knowledge of algebra.

The basic text is designed for a single-semester, sophomore- or junior-level course. It encompasses two-way analysis of variance to provide a firm foundation for an advanced or graduate-level course in statistical methods. The supplementary sections permit the inclusion of optional topics, which are not necessary to the sequence of the chapters and may be included or omitted as the interests or needs of the students dictate.

Introductory courses in educational statistics can be based on Chapters 1 to 9 (including topics through the t test) and Supplements C and G on test construction and data presentation. Advanced courses can use the initial chapters as review and then cover analysis-of-variance procedures and experimental design.

The outline inside the covers serves as a guide to a worked example and summary outline for each measure, permitting the text to serve as a handbook of basic statistical procedures. The function is aided by the inclusion of a large number of frequently used tables, some of which are in a revised and simplified form and others of which are not found in other texts.

Several topics are marked with single or double daggers († or ††). The single dagger denotes material based on procedures or concepts that are developed in later chapters, and these sections can be covered easily and quickly at the appropriate time. Topics marked with a double dagger are optional and can be by-passed without affecting the student's comprehension of the remaining material.

After studying the material in this text, the successful (i.e., surviving) student should be able to take data from a study, select and calculate the appropriate measures and tests, and finally, arrive at a reasonable description or interpretation of the situation. In more general terms, the student will appreciate some of the basic problems of analyzing and interpreting observations of the world about him, even if he never again has need for formal statistical procedures.

I wish to express my appreciation to the authors and publishers who permitted the use here of many statistical tables and, in particular, to the Literary Executor of the late Sir Ronald A. Fisher, F.R.S., to Dr. Frank Yates, F.R.S., and to Oliver & Boyd, Edinburgh, for permission to reprint Tables III, IV, V, and VI from their book *Statistical Tables for Biological, Agricultural and Medical Research.*

I thank William Hughes for his critical reading of the early drafts of the book and Cynthia Harris for her very fine editing of the final manuscript.

Herbert Friedman

Williamsburg, Virginia

contents

introduction
to
statistics

chapter one

HYPOTHESIS TESTING

A stranger approaches you at a party, produces a coin, and offers to play a simple game: He will toss the coin and if it lands heads, you will win two dollars ($2); if it lands tails, you will lose only one dollar ($1). The game is attractive, but you hesitate—the incessant smiling of the stranger has made you suspicious. Perhaps, you should first check the coin.

A reasonable test, it seems, might be to determine the percentage of times the coin lands heads in a test session. But this will not always work; suppose the coin lands heads 80 percent of the time. A perfectly fair coin may well land heads 4 out of 5 tosses by chance, but it would not be likely to land heads 20 times out of 25 tosses. Therefore, you should be concerned not only with the percentage of heads and tails, but also with whether or not the coin behaves as if it *might* be a fair coin. A fair coin tossed 20 times would not always land exactly 10 heads and 10 tails; 11, 12, or 13 heads or tails would be fairly common. The problem, therefore, reduces to the question: Does the coin that you are investigating land in a way consistent with its being a fair coin? In other words, would a fair coin behave as the coin being tested?

At first glance, it may seem illogical to test whether or not the coin can be assumed to be a fair one when you are really concerned with the possibility that the coin is unfair. Wouldn't it be easier just to test whether or not the coin is *unfair*? The main problem here is that there is no easy way to test directly whether or not a coin is unfair. Two examples will make this point clear:

1. We wish to check a coin to see if it might be unfair. The coin is tossed 3 times and lands heads. Do we conclude that the coin is unfair? No, since a fair coin might well land the same way 3 times in a row. The tossing is continued, and the total becomes 7 heads out of 8 tosses. Now do we consider

3

the coin to be unfair? No. Once again, a fair coin would do this often enough, though now we are definitely suspicious. With further testing the coin comes up 13 heads out of 15 tosses. We are now reasonably confident that a fair coin would not be so consistent, and therefore, this does not appear to be just a chance outcome that would occur with a fair coin. If the coin does not act like a fair coin, we can assume it is *not fair* and say that it is unfair, or *biased*. But we do not prove that it is unfair. We end up assuming that a fair coin would not behave as the one being tested.

This reasoning also holds for the opposite situation: If we toss a coin 8 times and get 4 heads, we cannot be sure the coin is fair, since a biased coin, which might tend to land heads three-fourths of the time, could, by chance, give the 4 out of 8 outcome. Once again, more data is needed; if the coin lands heads 250 times when tossed 500 times, we can be sure that the coin is *virtually* a fair coin, if not *exactly* a fair one.

2. Imagine a scene in which a police investigator enters the plush apartment of a principal suspect and finds him lying on the floor amid signs of a great struggle. Is the suspect dead? How does the detective find out? Is the suspect moving? No. Is the suspect breathing? No. Is the suspect cold? Yes. Is the suspect's heart beating? No. The suspect, now called the victim, appears to be dead. Note, however, that at no point could the policeman directly test the individual for being dead. All that he could do was test whether or not the suspect was alive. When the individual failed all the tests indicating life—that is, when his behavior, or lack thereof, did not appear to be within the range of living behavior—then the investigator could conclude that the man was not living, that he was dead. Similarly, if a coin does not act like a fair coin, then it is an unfair coin.

NULL HYPOTHESIS

One of the basic statistical procedures is the testing of the hypothesis that an observed outcome can be considered to be due to chance. This hypothesis of "nothing but chance" effects, which is stated more precisely in specific contexts, is known as the **null hypothesis** (abbreviated H_0). For the coin-tossing situation the null hypothesis is that the coin is a fair one and any deviation from 50 percent heads is due to chance, or random factors.

When we get results that would be very unlikely if the null hypothesis were true, we decide to act *as if* the null hypothesis is not true and conclude that the results are caused by something other than chance—that is, the coin is biased. We cannot *prove* that the null hypothesis is untrue, since there is always a finite possibility, however small, that it is true. Similarly, if the data are consistent with the null hypothesis, we act *as if* the null hypothesis is true, though we cannot prove that it is absolutely true.

The likelihood of the null hypothesis being true is one portion of the information needed to assess the fairness of a coin. If we decide the coin is unfair, we also have to know just how unfair the coin is. We would be reluctant to play with a coin that came up heads 6 times out of 7 (even though it might well be a fair coin), since it appears to be unfair. On the other hand, if the coin required 1,000 tosses before we could decide it was unfair (e.g., 531 heads out of 1,000 tosses), we could conclude that, although it is not a fair coin, it is *almost* fair and negligibly biased for most purposes. An average or typical coin may, in fact, be biased to this degree.

Thus there are three alternative conclusions when testing the null hypothesis as it applies to a coin:

1. We can decide to act as if the coin is unfair; that is, we will not play with the coin.
2. We can decide to act as if the coin is essentially a fair one; that is, we will play with the coin.
3. We can decide more testing is needed, for the coin looks suspicious, but not convincingly so.

Therefore, we need three items of information concerning the coin: The first is the likelihood that a fair coin would behave the same way as the test coin; the second is how biased the coin is; and the third is the direction of bias of the coin.

LEVEL OF SIGNIFICANCE

If the coin being tested behaves in a way inconsistent with its being a fair coin, we assume that it is a biased coin. In other words, if it is unlikely that a fair coin would give the results obtained, the null hypothesis (that the coin is fair) is not tenable, and we *assume* the coin is unfair.

How unlikely must an outcome be before we *reject* the null hypothesis? For the moment, we will arbitrarily set the following criterion: If a fair coin would not give the same results as the test coin more than once in 20 tests, we will assume the coin is unfair. That is, the test coin is **significantly** unlike a fair coin if the likelihood, or **probability**, of a fair coin behaving as the test coin is no more than 1 in 20 or 5 in 100. That is, probability $= p = 5/100 = .05$, or $p < .05$.* In scientific writing "significant" is a technical term referring to the rejection of the null hypothesis and is not a synonym for "important." Thus, results are regarded as being significant, or due to something other than chance, when the probability, or likelihood, of the outcome being a chance effect has been calculated and found to be sufficiently low (i.e., $p \leq .05$). The choice of an appropriate criterion, or **level of significance**, is discussed further in Chapter 3.

* The symbol $>$ means "is greater than" and $<$ means "is less than"; so $p < .05$ means "the probability value is less than .05 or 5 percent."

BINOMIAL TEST

The first problem in deciding if a coin is unfair is to calculate the probability values associated with tossing a fair coin. When there are two equally likely outcomes (probability of event p [heads] = probability of event q [tails] = 1/2), the probabilities for a specific series of events can readily be calculated by using the **binomial test**. The test is based on the fact that the tossing of a coin N times can be expressed algebraically in terms of the binomial expansion as $(H + T)^N$ or $(1/2 + 1/2)^N$. If the coin is tossed twice, the term becomes $(H + T)^2$ or $(1/2 + 1/2)^2$. A summary of the possible outcomes when a coin is tossed 1, 2, 3, and 4 times is given in Table 1.1. From this table it is clear that the likelihood of obtaining 3 heads (HHH) when a coin is tossed 3 times is 1 out of 8; that is, out of the 8 possible outcomes, each of which is equally likely, one possibility is 3 heads. The probability, therefore, of receiving 3 heads when a coin is tossed 3 times is 1/8 or .125.

TABLE 1.1

Number of Tosses, N	Outcomes, $(H + T)^N$	Number of Possible Outcomes
1	H, T	2
2	HH, HT, TH, TT	4
3	HHH, HHT, HTH, THH, HTT, THT, TTH, TTT	8
4	HHHH, HHHT, HHTH ... HHTT ... THTT, HTTT, TTTT	16

The probabilities for any coin-tossing outcome can be calculated on the basis of the binomial expansion, but the specific procedure is beyond the scope of this text. For convenience, the probability values for 5 to 50 tosses or events are given in Table H. The values listed in Table H refer to outcomes in either direction; that is, the likelihood of getting 9 or more heads *or* 9 or more tails when a fair coin is tossed 10 times is less than 5 percent or less than 5 times in 100.* The probability value in Table H for 9 heads is actually the sum of two probabilities—that for exactly 9 heads and that for exactly 10 heads. Thus the value given is the probability for 9 or more heads. This point is not obvious, nor is the beginner initially comfortable with this approach; but a brief example may help: A fair coin tossed 1,000 times is very unlikely to land heads *exactly* 500 times, but it is likely (a 50 percent chance) to land heads *at least* 500 times (i.e., 500 or more times). The material in subsequent chapters will clarify this point further.

Table H also makes it easy to see what happens to the probability value for a given proportion of heads and tails as the sample size (i.e., number of

* Probability values relating to outcomes in either direction are called **2-tailed** probability values and will be discussed later in this text.

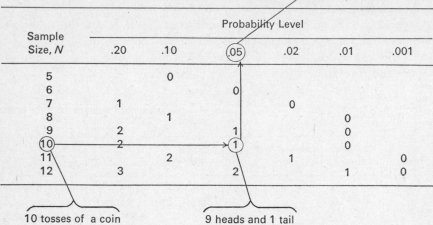

probability level associated with
1 tail (or 9 heads) in 10 tosses

TABLE 1.2. Each Table Entry Indicates the *Maximum* Number of Times the *Less* Frequent of the Two Events May Occur for Each Probability Level.

Sample Size, N	Probability Level					
	.20	.10	.05	.02	.01	.001
5		0				
6			0			
7	1			0		
8		1			0	
9	2		1		0	
10	2		1		0	
11		2		1		0
12	3		2		1	0

10 tosses of a coin   9 heads and 1 tail

tosses) is increased. For example, with 100 percent heads (0 tails) the probability value decreases rapidly as N increases from 5 to 11—as shown in Table 1.2, which gives a section of Table H. Similarly, the p value for 80 percent heads decreases as the sample size increases, as shown in Table 1.3. A given outcome (other than 50 percent heads and 50 percent tails), then, becomes increasingly unlikely as the sample size is increased. In other words, 4 heads out of 5 tosses is not unreasonable for a fair coin, but 12 out of 15 and 16 out of 20 are clearly unlikely—though still possible for a fair coin. As the number of coin tosses increases, even small deviations from its being fair become statistically significant (e.g., when $N = 50$, 33 heads is significant with $p < .05$).

TABLE 1.3

N	Number of Heads	Percentage of Heads	Probability, p
5	4	80	> .20
10	8	80	< .20
15	12	80	< .05
20	16	80	< .02
25	20	80	< .01
35	28	80	< .001

SIZE OF THE EFFECT

Now that we can determine the probability of a fair coin acting like the coin being tested, we need a measure of how biased the coin appears to be. An obvious and suitable measure is the percentage of heads or tails (whichever is larger). A fair coin would land with approximately 50 percent heads, whereas a totally biased coin would land with 100 percent heads or 100 percent tails. Slightly and moderately biased coins would be between these extremes. In the next chapter the measurement of the size of the effect will be discussed in more detail.

To return to the problem raised at the start of the chapter, assume that we have tested the offered coin and calculated the number of heads and tails, the probability value, and the size of the effect. A summary of reasonable conclusions for 4 different outcomes is presented in Table 1.4.

TABLE 1.4. "Heads You Win, Tails You Lose"—Summary of Conclusions for Four Coin Tests.

Example	Number of Heads	Number of Tails	p	Size of Effect, %	Conclusion
A	2	12	$< .05$	86 (moderate)	Do not play—the coin is against you. (Act as if the coin is unfair.)
B	0	5	$< .10$	100 (large)	Be very wary. (Coin is suspect but needs more testing.)
C	8	1	$< .05$	89 (large)	Play—it is to your favor. (Act as if the coin is unfair.)
D	9	11	$> .20$	55 (small)	Play—fair enough. (Act as if the coin is fair.)

The probability, p, value provides a measure of the *likelihood* that the coin is fair; the size of the effect is a measure of the *degree* to which the coin is biased, and the ratio of heads to tails indicates the *direction* of the coin's bias. In example A of Table 1.4, the coin is significantly ($p < .05$) and moderately biased in favor of tails. We would be wise to substitute another coin. A similar situation is seen in example C, but now the bias is in our favor. It would be profitable, if unsporting, to use the coin. In B, the coin, although not significantly biased in favor of tails ($p > .05$), is certainly not one we would be willing to use without further testing. In D, the coin is fair enough to use, even though we do not know if it is perfectly fair.

† NORMAL DISTRIBUTION APPROXIMATION FOR THE BINOMIAL TEST WHEN $N > 50$ OR $p \neq q$

Table H for the binomial test can be used only when $p = q = 1/2$ and when the sample size (N) is no more than 50. As N gets larger, the distribution for the binomial approaches a normal distribution in shape, and the probabilities associated with the normal distribution provide a good approximation of the probabilities for the binomial. All we need to determine the p value is the appropriate Z score, which is obtained with the following formula:

$$Z = \frac{|X - Np| - 1/2}{\sqrt{N(p)(q)}} \tag{1.1}$$

where X = number of occurrences of event A
N = total number of observations or events
p = probability of event A
q = probability of event B = $1 - p$
$|\ \ |$ = absolute value of the enclosed term; negative values are treated as if positive.

For example, 42 heads are obtained when a coin is tossed 64 times. The observed value (X) is 42. The value that we would expect by chance (Np) is $64 \times 1/2$ or 32. The 1/2 that we subtract from the absolute value of this difference is a correction for continuity (equivalent to that used for χ^2) and improves the match of the normal distribution to the distribution of the binomial. The denominator for this example works out to be $\sqrt{16} = 4$. In Table D for the normal distribution a Z score of 2.37 gives a probability value of .0089 for a one-tailed test, and this value is doubled for a two-tailed test to equal approximately .018. Therefore, the probability of getting 42 or more heads when tossing a coin 64 times is $< .02$, and we would reject the null hypothesis that the coin is fair. The coin is significantly ($p < .02$) but only slightly ($r_m < .3$ from either Table B or C) biased in favor of heads.

This approximation is also useful when $p \neq q$. A good rule of thumb is that the normal distribution approximation is satisfactory when the sample size (N) is large enough so that the value of $N(p)(q) \geq 5$. In the case of coin tossing, this would require that $N > 20$ (i.e., $20 \times 1/2 \times 1/2 = 5$), whereas in a die- (one of a pair of dice) tossing study where $p = 1/6$, $q = 5/6$, we would need an $N \geq 36$ (i.e., $36 \times 1/6 \times 5/6 = 5$) for an adequate approximation. When $N(p)(q)$ is less than 5, probability values should be calculated exactly from the binomial expansion. The process is cumbersome, and the procedure is given in many standard references (e.g., Siegel 1956).

BINOMIAL TEST

I   Using Table H when both events are equally likely ($p = q$) and there are no more than 50 trials ($N \leq 50$):

Formula X = number of times the least frequent event occurs.

Table H   provides p values based on X and N.

Table B   provides r_m values based on p and N.

Procedure
1. Count the number of times each of the two events occurs.
2. X = the number of times the least frequent event occurs.
3. Use Table H to determine p on the basis of X and N.
4. Use Table B to determine r_m on the basis of p and N.

Example   In a test for fairness a coin is tossed 20 times. It lands heads up 17 times and tails up 3 times.
$X = 3$ = smaller of the two outcomes (17 and 3). From Table H, $p < .01$ with $X = 3$ and $N = 20$. From Table B, $r_m > .54$ with $p < .01$ and $N = 20$.

Conclusion   The coin is significantly ($p < .01$) and appreciably biased ($r_m > .54$) and tends to land heads up.

II   Normal Distribution Approximation: When both events are not equally likely ($p \neq q$) or when $N > 50$ and $p = q$, the normal distribution approximation of the binomial distribution can be used. The probability value (p) is given by the area of the normal distribution lying beyond point Z (i.e., in the tails of the distribution), with Z based on X, p, q, and N.

Formula $$Z = \frac{|X - Np| - 1/2}{\sqrt{N(p)(q)}}$$

when $N(p)(q) \geq 5$

Table D	provides p values based on Z.

Table D provides p values based on Z.
Note: The table entries should be doubled to obtain two-tailed probability values.

Table B provides r_m values based on p and N.

Restrictions The approximation is suitably accurate when $N(p)(q) = 5$ and increases in accuracy as N gets larger. When $p = q = 1/2$, $N = 20$ is sufficient. As p and q diverge, larger sample sizes are needed. The accuracy of the approximation is improved by the correction for continuity, which involves the subtraction of 1/2 from the absolute value of the $(X - Np)$ difference. When $p \neq q$ and $(p)(q) < 20$, use the binomial expansion directly, as discussed in Siegel (1956).

Procedure
1. Check that $N(p)(q) \geq 5$.
2. Count the number of times each of the two events occurs.
3. $X =$ the number of occurrences of the event associated with p.
4. Calculate Z based on X, N, p, and q.
5. Use Table D to determine the p value based on Z.
6. Use Table B to determine the r_m value based on p and N.

Example A die is tested for a tendency to land with a six. It is tossed 60 times and lands with a six 17 times and some other number 43 times. For six $p = 1/6$; for numbers other than six $q = 5/6$.

$$N(p)(q) = 60 \ (1/6)(5/6) = \frac{60(5)}{36} = 8.3333 > 5$$

$$Z = \frac{|17 - 60(1/6)| - 1/2}{\sqrt{60(1/6)(5/6)}} = \frac{|17 - 10| - 1/2}{\sqrt{8.3333}}$$

$$= \frac{6.5}{2.89} = 2.25$$

From Table D, $p < .05$ with $Z = 2.25$.
From Table B, $r_m > .25$ with $p < .05$ and $N = 60$.

Conclusion There is a significant ($p < .05$) tendency for the die to land with a six, but the size of the effect is small ($r_m > .25$).

SUMMARY

1. *Hypothesis testing* involves the use of statistical measures of data to test if certain hypotheses, or descriptions of the data, are reasonable.

2. The *null hypothesis* (H_0) is the assumption that any difference between an observed outcome and an expected outcome is due to chance or random factors. When the difference between observed and expected outcomes is too large, H_0 is not tenable and is *rejected*.

3. The *level of significance* is a probability value (p) that is arbitrarily set as being significant. If the likelihood of an observed outcome occurring by chance (if H_0 were correct) is less than the significant p value, H_0 is *assumed* to be incorrect and is rejected.

4. The *binomial test* is used for situations with two possible outcomes (e.g., heads/tails) and sample sizes of 50 or less. When the two outcomes are equally likely (i.e., $p = q$), the appropriate probability values can be readily obtained from Table H.

5. The *size of the effect* is a measure of the degree of bias or the difference between observed and expected outcomes.

†6. The *normal distribution approximation for the binomial test* is used when $p = q$ (since $N > 50$) or when the two outcomes are not equally likely ($p \neq q$) and the sample size is sufficiently large so that $N(p)(q) \geq 5$.

chapter two

Consider again the example of the coin-tossing problem, which was analyzed by the binomial test in the previous chapter. This type of problem can also be analyzed by a more general approach, which is applicable to many situations for which the binomial test is unsuited. With this approach we obtain a numerical measure of the degree to which the observed outcome differs from that which would be expected on the basis of chance. Consider a coin that lands 20 heads and 4 tails out of 24 tosses, as illustrated in Figure

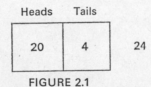

Heads Tails

| 20 | 4 | 24 |

FIGURE 2.1

2.1. For a fair coin the expected outcome would be 12 and 12, since the chance outcome for a fair coin is an equal number of heads and tails. If the "true" behavior of the coin is fair, is it likely that out of 24 tosses the coin would land heads 20 times and tails 4? Before we can deal with this question, certain basic terms must be defined.

POPULATIONS AND SAMPLES, STATISTICS AND PARAMETERS

In a study of possible differences between men and women with regard to a political question, we would generally select a group of men and a group of women and obtain their views. Obviously, we could not expect to test *every* man and *every* woman. This would be practically impossible in terms of time, effort, and money required and logically impossible in that at any given instant individuals enter and leave any specific group of men and women.

13

The entire group of individuals or cases comprising the subject of an investigation is known as the **population**. A population can also be defined as all the individuals or cases that fit into a certain category. All students at a given university are a population; all C57BL/6J mice are another population; and the population of all mammals includes both of these smaller populations. Populations can be any size—ranging from 1 (e.g., all first Presidents of the United States) to infinity (e.g., all even numbers).

Since working with total populations is not practical, it is necessary to select a group of individuals, or subjects, from the population. This group of subjects is known as a **sample**. In order for the sample to give a useful picture of the population, it must be **representative** of the population. In other words, it must appropriately reflect the characteristics of the population. A sample that is not representative, one that gives a distorted picture of the population from which it is drawn, is said to be **biased**. When a sample is tested and measured, the measure of the sample is called a **statistic**. The measure of a population is called a **parameter** and is based on the testing of each member. The degree to which the sample is an adequate reflection (i.e., representative) of the population determines the degree to which the statistic is an adequate estimate of the parameter. Therefore, the data from a good sample can be very informative about, or generalized to, the population from which the sample is drawn. The relationship among the terms population, sample, statistic, and parameter is summarized in Figure 2.2.

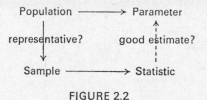

FIGURE 2.2

REPRESENTATIVE SAMPLES

We must be careful when deciding whether or not a sample is representative, for the "representativeness" of a sample can be determined *only* with regard to the characteristics that are being measured. For example, basketball players are certainly not representative of college students in terms of height, though they may be adequately representative with regard to learning ability, attitudes, or liver metabolism. The two most popular subjects for psychological research, the college sophomore and the white rat, have been criticized as not being an adequate sample of *people*. The typical sample of college students tends to be restricted in age range, as well as being from a higher socioeconomic background, more healthy, more motivated, and (presumably) more intelligent than a random sample of the nation's population. If these variables are critical to a study, the conclusion drawn from a study on college students cannot be reasonably generalized to the total population, although it can be generalized to a larger population of students. On the other hand, in studies of simple learning processes which are assumed to be general throughout a wide range of vertebrates, the rat may be a suitable and convenient as well as representative subject.

RANDOM SAMPLES

The most common way of selecting a representative sample of subjects is to use **random sampling**. With this procedure each individual in the population has an equal chance of being included in the sample, for the selection process is random or, at least, nonsystematic. Although the samples may differ by chance from one another and from the population, we can be confident (even without knowing the characteristics of the population) that they show no consistent bias. Many standard statistical tests are based on the assumption that the data come from randomly selected samples.

THE χ^2 STATISTIC

We now return to the original problem of the coin that landed with 20 heads out of 24 tosses. The appropriate statistic is a numerical measure of the difference between the values actually observed (O) and those expected (E) if H_0 were true and the coin was really fair. These values can be conveniently shown in a box or *cell* for each possible outcome, as illustrated in Figure 2.3. It is tempting to calculate the difference between the O and E values for both heads and tails and add the results:

$$\sum (O - E) = (20 - 12) + (4 - 12) = (+8) + (-8) = 0^* \qquad (2.1)$$

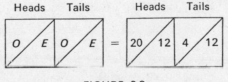

FIGURE 2.3

However, this clearly does not give the proper statistic, since the sum is always 0. The minus and plus differences would not cancel out if the absolute value of the differences were used, but this is not algebraically satisfactory. A suitable alternative is to square the differences (since $-8^2 = +64$) so that all

---

* Uppercase sigma ($\sum$) in the formula means "sum of"; thus $\sum (O - E)$ means the sum of the differences between O and E values.

the terms to be added become positive:

$$\sum (O - E)^2 = (20 - 12)^2 + (4 - 12)^2 = (8)^2 + (-8)^2 = 64 + 64 = 128 \tag{2.2}$$

So far, so good; the value of the term $\sum (O - E)^2$ will get larger as the differences between the observed and expected values grow larger. But consider the situation shown in Figure 2.4. In this case, the $(O - E)$ differences appear to be relatively small. However, the value of $\sum (O - E)^2$ is the same as in the example of 20 heads out of 24 coin tosses. Therefore, we have to take into account not only the size of the $(O - E)$ differences but also the size of the expected value for each outcome or *cell* (which increases as the total number of tosses increases), and this can be done in the following fashion:

$$\sum \left[\frac{(O - E)^2}{E} \right] = \chi^2 \tag{2.3}$$

$$\frac{(20 - 12)^2}{12} + \frac{(4 - 12)^2}{12} = \frac{64}{12} + \frac{64}{12} = \frac{128}{12} = 10.67 \tag{2.4}$$

The χ^2 value (called **chi square**) is the statistic we want. It appropriately measures the differences between the observed and expected values and takes the absolute size of the expected values into account. But just what does the χ^2 value mean?

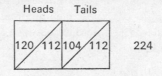

FIGURE 2.4

SAMPLING DISTRIBUTIONS FOR χ^2

In order to determine the likelihood of the observed outcome ($\chi^2 = 10.67$) occurring with a fair coin, we need a picture of all of the possible χ^2 values that could arise by chance. We could obtain this picture by tossing a fair coin 24 times, calculating the χ^2 value for the outcome, and repeating this procedure hundreds or thousands of times. The resulting distribution of χ^2 values would give the desired picture of the outcomes to be expected by chance. Such a χ^2 distribution for random samples is known as a **sampling distribution**. In the coin-tossing problem, most of the χ^2 values will be 0 or close to 0. Larger values will occur less frequently. Once we know the shape or properties of the sampling distribution, we can determine easily if the observed outcome is likely to have occurred by chance. If the observed value of χ^2 is greater than the values reasonably expected to occur by chance, then we conclude that, since the coin did not behave as a fair coin is expected to behave, the coin is unfair. On the other hand, if the difference between the observed and expected values is small, then the observed outcome could reasonably be obtained with a fair coin. The characteristics of the sampling distribution for χ^2 have been calculated mathematically and are conveniently given in the form shown in Table F.*

DEGREES OF FREEDOM

Although the χ^2 statistic as defined in formula (2.3) is a suitable measure of the difference between the observed and expected values, it does not take into account the number of categories or cells involved. In problems based on tossing a die or measuring the frequency of purchase of four makes of automobiles, there would be more than two cells, as shown in Figure 2.5.

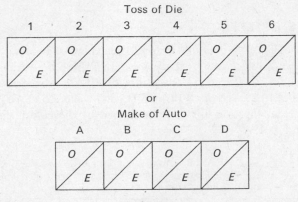

FIGURE 2.5

Under these conditions the number of cells will affect the χ^2 value. Given small, chance differences between observed and expected values for each cell, the final value of χ^2 will increase as the number of cells increases. Therefore, there is a different sampling distribution for each separate situation in terms of the number of cells. The number of cells is not used directly, but rather in terms of a measure called **degrees of freedom (df)**. The concept of degrees of freedom, which will be discussed again with regard to the standard deviation in Chapter 7, refers to the number of measurements that are free to vary and, therefore, that convey information. In the case of χ^2 we must assume that the *marginal frequency*, the total number of observations, is a fixed value. If in a simple coin-tossing situation there are 15 heads out of 20 tosses, there must be 5 tails. Given a fixed total number of observations, all of the information is provided in one of the two cells. Knowing the number of heads, we can

* Table H, giving probability values associated with the binomial test, is similarly based on the sampling distribution for the binomial when $p = q = 1/2$.

calculate the number of tails. In the case of another example, the sales figures for four types of automobiles, if we know the total number of automobiles sold and the number sold in the first three categories, then we can determine the sales of the fourth type. The degrees of freedom, then, equals the number of cells minus one:

$$df = C - 1 \tag{2.5}$$

where C = number of cells.

Note that Table F is organized in terms of the degrees of freedom and the value obtained for χ^2. At this point, it might seem easier to write the table in terms of χ^2 and the number of cells. However, when the χ^2 values are for more than one sample, the formula for calculating df is slightly different, and df is no longer equal to $C - 1$, even though the sampling distributions are still functions of df.

MAGNITUDE OF EFFECT

Back to the problem of the coin that lands 20 heads out of 24 tosses, for which the calculated χ^2 value is 10.67. If we look at Table 2.1—which gives the appropriate section of Table F—the value of 10.67 is between the table values of 6.64 and 10.83 with $df = 1$ (the first row). Therefore, the likelihood, or probability, that this is a fair coin (indicated by the column heading) is between .01 and .001.

TABLE 2.1

df	Probability					
	.20	.10	.05	.02	.01	.001
1	1.64	2.71	3.84	5.41	6.64	10.83
2	3.22	4.60	5.99	7.82	9.21	13.82
3	4.64	6.25	7.82	9.84	11.34	16.27
4	5.99	7.78	9.49	11.67	13.28	18.46
5	7.29	9.24	11.07	13.39	15.09	20.52

$.01 > p > .001$

table χ^2 values

$df = 1$

On the basis of the calculated probability value, we decide the coin is unfair; that is, it is significantly biased. How biased is the coin? In Chapter 1 we used the percentage of heads and tails to measure the size of the bias. We can now consider a more general measure of bias, which is also suitable for situations more complex than coin tossing. This measure is of the relative difference between the observed outcome (O values) and the outcome expected if the null hypothesis were true (E values).

With the binomial test we found that as the sample size increases for a

TABLE 2.2. Magnitude of Experimental Effect (r_m) as a Function of the Sample Size Needed to Detect the Effect at a Given 2-Tailed Probability Level

Sample Size	Probability				
	.20	.10	.05	.01	.001
1	.95	.99	1.00	1.00	1.00
2	.80	.90	.96	.99	1.00
3	.69	.81	.88	.96	.99
4	.61	.73	.81	.92	.97
5	.55	.67	.75	.87	.95
20	.28	.36	.42	.54	.65
21	.28	.35	.41	.53	.64
22	.27	.34	.40	.52	.63
23	.27	.34	.40	.51	.62
24 →	.26	.33	.39	.50	.61
25	.25	.32	.38	.49	.60

$N = 24$ $p = < .01$ $r_m > .50$
for $p < .01$
and $N = 24$

given effect (e.g., proportion of heads), the p value decreases. The situation is the same for the χ^2 test. We can, therefore, use the probability value and sample size for a test on a given set of data to provide a measure of the size of the effect, or bias. This type of measure is particularly useful, since it is suited to both the χ^2 and binomial tests and to the other tests considered in this text. This measure, which is called the *magnitude of effect* and abbreviated r_m, is discussed more fully in Chapter 12. Values of r_m for a wide range of sample sizes and probability values (p) are given in Table B. The use of Table B to obtain r_m values from the p value and sample size, which for the binomial and χ^2 tests is N, is illustrated in Table 2.2.

The r_m value as a measure of the size of the observed effect is an inverse function of the number of observations necessary to detect the effect as being significant. If a coin is so highly biased that it always lands heads, we could detect this coin as being unfair with only 6 tosses. For this coin the r_m is large ($> .71$). An almost, but not quite, fair coin may be tossed 120 or more times before it is seen as being biased, and for this coin the r_m is small ($< .18$). Therefore, as the bias increases, the effect increases, and fewer observations are necessary to detect the outcome as being a non-chance one. Note that when conducting a study, we must take enough observations (in this case, coin tosses) to be able to detect the smallest important effect, or bias. Table B indicates that, if we are interested in a large bias only (e.g., 80 percent heads), the coin should be tossed at least 14 times.

The upper half of Table 2.3 shows the effect upon r_m when the sample size is increased and the size of the effect is constant (80 percent heads). The χ^2 value increases and the probability value decreases, but the r_m value, reflecting the proportion of heads, remains constant. The lower half of Table 2.3 shows how the proportion of heads and the r_m value decrease when the same χ^2 (and p value) is obtained with samples of increasing size. In other words, as the sample required to produce a significant χ^2 value increases, the r_m value decreases.

TABLE 2.3

Heads	Tails	Percentage of Heads	N	χ^2	p	r_m
4	1	80	5	1.80	< .20	> .55
8	2	80	10	3.60	< .10	> .50
16	4	80	20	7.20	< .01	> .55
32	8	80	40	14.40	< .001	> .49
11	3	79	14	4.57	< .05	> .50
18	7	72	25	4.16	< .05	> .38
38	22	63	60	4.27	< .05	> .25
71	49	59	120	4.03	< .05	> .18

Values of r_m range from .00, for outcomes exactly equal to that expected by chance, and approach 1.00 for very large effects. The use and interpretation of measures of magnitude of effect will be discussed more fully in Chapter 12. In order to use the r_m measure early in the text, we will arbitrarily treat values in the following manner:

.00 → .30   small effect, close to the chance outcome expected under H_0
.30 → .50   moderate effect
.50 +   large effect, results highly different from those expected under H_0

Actually, the interpretation of an obtained r_m value as being large or small is always a function of the specific situation being measured. A given r_m value may reflect an acceptable or important outcome in one study and a small effect in a different situation.

To return to the coin-tossing example, Table B indicates that a probability value of $< .01$ obtained with 24 observations ($N = 24$) represents an r_m value of $> .50$. A reasonable interpretation is that the coin is significantly ($p < .01$) and appreciably ($r_m > .50$) biased in favor of landing heads.

CORRECTION FOR CONTINUITY

Applying the binomial test to this situation where the coin is tossed 24 times would lead us to essentially the same conclusion. There would be a slight difference in p values, but this could be eliminated by a minor correction in the calculation of χ^2. This adjustment is used when $df = 1$ (i.e., when there are 2 cells) and is necessary because of the nature of category data. For example, when tossing a coin 10 times, we could get 2 or 7 heads, but not 2.5 or 7.3 heads; with 100 tosses, 25 or 73 heads would be possible. The sampling distribution for χ^2 is based on an infinite sample size, and the stepwise nature of the data, when N is small, leads to a slight but consistent mismatch. When $df = 1$, this mismatch causes slightly incorrect probability values. The appropriate correction is called the **correction for continuity** and involves the subtraction of 1/2 from each $(O - E)$ difference. Thus

$$\chi_c^2 = \sum \frac{(|O - E| - 1/2)^2}{E} \tag{2.6}$$

where $\chi_c^2 = \chi^2$ corrected for continuity

$(|O - E|)$ means the absolute value of $(O - E)$, so the difference is always reduced by the correction.

For the example the calculation works out like this:

$$\chi_c^2 = \frac{(|20 - 12| - 1/2)^2}{12} + \frac{(|4 - 12| - 1/2)^2}{12} = \frac{(7.5)^2}{12} + \frac{(7.5)^2}{12} = 9.38 \tag{2.7}$$

Thus the answer is 9.38 instead of 10.67, as obtained with equation (2.4). As the sample size increases, the $(O - E)$ differences also tend to increase, and the relative size of the correction decreases. Adjusting $(4)^2$ to $(3.5)^2$ involves a larger correction than changing $(20)^2$ to $(19.5)^2$.

Restrictions on the use of the χ^2 test are discussed in the following chapter.

SINGLE-SAMPLE TESTS

Both the binomial test (Chapter 1) and the χ^2 single-sample test considered here deal with a situation where there is only one sample. The data obtained from the sample are tested to determine if the sample can reasonably be considered a random sample from a given population. In order to do this we must have some previous knowledge of or theory about the characteristics of the population. In the case of testing a coin for fairness, we must know that the expected or chance outcome would be an equal number of heads and tails. With the binomial test these expected values are the basis for the calculation of the probabilities of the various outcomes. In the case of the χ^2 test the expected (E) values are stated explicitly. Consider the following two examples:

1. At a certain university 2/3 of the students are men and 1/3 are women. You are interested in a possible selection factor in enrollment for a course in Advanced Desserts offered by the Home Economics Department. The distribution of the 36 students in the course is 2 men and 34 women, as

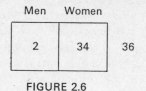

FIGURE 2.6

shown in Figure 2.6. If the students in this course were a random selection of the university population, a 2:1 proportion of men to women or 24 men and 12 women in the course would be expected (see Figure 2.7). You calculate $\chi_c{}^2$:

$$\chi_c{}^2 = \frac{(|2 - 24| - 1/2)^2}{24} + \frac{(|34 - 12| - 1/2)^2}{12}$$

$$= \frac{(21.5)^2}{24} + \frac{(21.5)^2}{12} = 57.78 \tag{2.8}$$

$$p < .001 \qquad r_m > .49$$

The actual enrollment in the course is significantly ($p < .001$) and appreciably ($r_m > .49$) different from the expected values, with the course tending to attract a greater proportion of women.

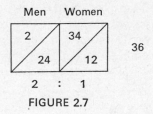

2 : 1

FIGURE 2.7

2. In a genetics experiment pink pea flowers are crossed. Classical theory predicts that in the next generation white, pink, and red flowers will appear in a proportion of 1:2:1. The actual findings for 80 offspring are shown in Figure 2.8. The expected number of white, pink, and red flowers is 20, 40, 20, respectively. χ^2 is calculated as follows:

$$\chi^2 = \frac{(16 - 20)^2}{20} + \frac{(42 - 40)^2}{40} + \frac{(22 - 20)^2}{20} = 1.1 \tag{2.9}$$

$$p > .20 \qquad r_m < .20$$

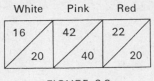

FIGURE 2.8

In this case, there is a nonsignificant ($p > .20$) and small ($r_m < .20$) difference between the observed and expected values, and we conclude that the result is consistent with the theory.

Note from these two examples that the expected values for a single-sample test come either from prior knowledge, as in example 1, or from some theoretical formulation, as in example 2.

GOODNESS-OF-FIT TESTS

The single-sample test deals with the similarity between the observed and expected values. This type of test is often called a **goodness-of-fit** test. The example of the genetics experiment is a typical case of testing to see if the observed data are consistent with what is expected. Goodness-of-fit tests are used to determine if a given sample could reasonably be assumed to come from a population having certain characteristics. Such tests can be important when assumptions are made concerning the population from which samples are drawn.

χ^2 SINGLE SAMPLE

Formula

$$\chi^2 = \sum \left[\frac{(O_j - E_j)^2}{E_j} \right]$$

$$\chi_c^2 = \sum \left[\frac{(|O_j - E_j| - 1/2)^2}{E_j} \right]$$

where O = observed frequency for event j
E = expected frequency for event j
df = degrees of freedom = number of cells $- 1$
N = number of cases = $\sum O$

Table F   provides p values based on χ^2 and df.

Table B   provides r_m values based on p and N.

Table C   provides r_m values based on χ^2 and N.

Restrictions   The expected values (E) are based on prior knowledge of the situation or on theoretical considerations.
    All observations should be independent, with no individual subject being tested more than once, so that $N = \sum O$ = number of different subjects tested.
    The expected values (E) for all cells should be greater than 5 so as to avoid inflated χ^2 values. No cell should have an E value of 0.
    With two cells ($df = 1$) the test is equivalent to the binomial test, and χ_c^2, incorporating the correction for continuity, should be used.

Procedure   1. From prior knowledge or theory determine the E value for each cell.
    2. Calculate the $(O - E)$ difference for each cell and square the term to obtain $(O - E)^2$. When $df = 1$,

subtract 1/2 from the absolute value of each $(O - E)$ difference and then square the term to obtain

$(|O - E| - 1/2)^2$.

3. For each cell divide the term from step 2 by the E value for that cell.
4. Add the values from step 3 for all cells to obtain χ^2 (or χ_c^2 for $df = 1$).
5. Use Table F to determine the p value based on χ^2 and df = number of cells $- 1$.
6. Determine the r_m value from Table B based on p and N or Table C based on χ^2 and N.

Example 1

In an experiment 24 college students are each given a questionnaire and then sent one at a time to fill it out. Each student may choose one of three adjacent class-rooms, two of which were originally empty, the third containing six "students" working busily. The distribution of the students among the three rooms is:

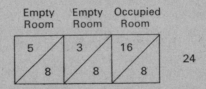

The E value for each cell is 8 and reflects the null hypothesis that there is no preference for any of the rooms.

df = number of cells $- 1 = 3 - 1 = 2$

$$\chi^2 = \frac{(5 - 8)^2}{8} + \frac{(3 - 8)^2}{8} + \frac{(16 - 8)^2}{8}$$

$$\chi^2 = \frac{9}{8} + \frac{25}{8} + \frac{64}{8} = \frac{98}{8} = 12.25$$

25

From Table F, $p < .01$ with $\chi^2 = 12.25$ and $df = 2$.
From Table B, $r_m > .50$ with $p < .01$ and $N = 24$.
From Table C, $r_m > .55$ with $\chi^2 = 12.25$ and $N = 24$.

Conclusion

There is a significant ($p < .01$) and appreciable ($r_m > .55$) difference in the proportion of students in each of the three rooms, with the students tending to use the room that was already occupied.

Example 2

The shell game at a local carnival employs the standard 3 shells. After a typical day of 30 players the number of winners and losers is:

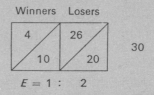

df = number of cells $- 1 = 2 - 1 = 1$

The correction for continuity should be used. The proportion of winners to losers would be $1 : 2$ if only chance factors were involved.

$$\chi_c^2 = \frac{(|4 - 10| - 1/2)^2}{10} + \frac{(|26 - 20| - 1/2)^2}{20}$$

$$\chi_c^2 = \frac{(5.5)^2}{10} + \frac{(5.5)^2}{20} = 4.5375$$

From Table F, $p < .05$ with $\chi^2 = 4.5375$ and $df = 1$.
From Table B, $r_m > .35$ with $p < .05$ and $N = 30$.
From Table C, $r_m > .35$ with $\chi^2 = 4.5375$ and $N = 30$.

Conclusion

The proportion of winners is significantly ($p < .05$) and moderately ($r_m > .35$) smaller than would be expected by chance.

SUMMARY

1. A *population* is the entire group of individuals, objects, or events sharing a particular characteristic. A *sample* is a portion of a population. A *statistic* is a measure of a sample, and a *parameter* is a measure of the population. A statistic approximates a parameter to the degree that the sample is *representative* of the population, that is, to the extent that it is not *biased*.

2. *Random sampling* is a commonly used procedure for obtaining suitably unbiased samples; the sample is selected on a chance, or random, basis and thus each member of the population has an equal opportunity to be selected.

3. The χ^2 *statistic* is a numerical measure of the difference between observed and expected values in a situation with two or more possible outcomes.

4. *Sampling distributions* show the distribution of chance outcomes for a statistic, when H_0 is correct, and are used to determine probability values for an obtained value of the statistic.

5. *Degrees of freedom* (df) is a measure of the number of scores (e.g., cell frequencies) in the data that provide information.

6. The *magnitude of effect* (r_m) is a measure of the *size* of an observed outcome or of the difference between observed and the expected outcomes. Both the r_m and probability (p) values are aids in interpreting the data.

7. The *correction for continuity* is made, when χ^2 values are calculated, by subtracting $1/2$ from the $(O - E)$ difference in each cell. The correction is necessary for accurate probability values when $df = 1$.

8. *Single-sample tests* are used to determine if a given sample could reasonably be assumed to come from a particular population. The characteristics of the population (i.e., E values) must be known in advance of the test.

9. *Goodness-of-fit tests* are a commonly used application of single-sample tests. The distribution of outcomes in a sample is compared to the distribution expected for a sample from a particular population.

chapter three

MULTI-SAMPLE χ^2 COMPARISONS

The previous chapters on the binomial and χ^2 tests discussed single-sample tests—that is, situations in which the obtained data were compared to some previously expected outcome. We are now concerned with the problem of comparing two samples with each other in order to decide whether we can assume they are essentially the same or different. For example, to test possible selective factors in the registration of men and women for two equivalent sections of a freshman English course, we note the number of male and female students registering in the section taught by the eligible new assistant professor Harry Cleftchin and the other section taught by that friendly graduate student Gwendelyn Tightsweater. The data look like that given in Figure 3.1. A casual inspection of the data suggests there are differences in the proportion of male and female students in the two sections, but a numerical measure (in terms of the χ^2 statistic) is needed to describe this difference. Having obtained the observed (O) values, we need only the expected (E) values.

	Male	Female	
H. C.	11	13	24
G. T.	10	2	12
	21	15	36

FIGURE 3.1

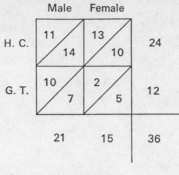

FIGURE 3.2

The null hypothesis in this case is that presumably both samples were drawn from the same population, their differences being due to chance only. If the null hypothesis is correct, what would be the best picture or estimate of the population from which the samples have been drawn? Clearly, the sum of the two samples would give the best picture. In this case the total of the two samples is 21 male and 15 female students. We can now obtain the expected values for each cell. Of the total group of students, 24 out of 36, or 2/3, are in Harry's class. Therefore, of the 21 male students we would expect 2/3, or 14, to be in his class and the remaining 1/3 to be in Gwendelyn's section. The same reasoning can be applied to the two cells for the female students, and the observed and expected values are shown in Figure 3.2.

The process of obtaining expected values is readily reduced to a simple mathematical formula. The expected value (E) for any cell (cell$_{r_i c_j}$)* equals the marginal frequency for the row $(\sum r_i)$ containing that cell times the marginal frequency for the column $(\sum c_j)$ containing that cell, divided by the total number of cases (N):

$$E_{r_i c_j} = \frac{(\sum r_i)(\sum c_j)}{N} \qquad (3.1)$$

The expected values in the example were obtained by exactly the same procedure as that stated in formula (3.1). Twenty-four of the 36 students (24/36) are in Harry's section, and we would expect that 24/36, or 2/3, of the 21 male students would be in Harry's section. In terms of formula (3.1),

$$\frac{24}{36} \times 21 = \frac{(24 \times 21)}{36} = 14 \qquad (3.2)$$

In contrast to the single-sample test, obtaining expected values here does

---

* Where r_i stands for a given row and c_j a given column, so that cell$_{r_i c_j}$ refers to the cell at the intersection of r_i and c_j in the χ^2 array.

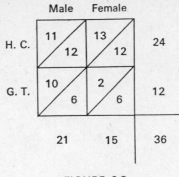

FIGURE 3.3

not require knowledge of the population from which the samples were drawn. The samples are compared to each other rather than separately to some prior expectation.

A frequent mistake is to use the (incorrect!) expected values shown in Figure 3.3. We just *know* that these *E* values cannot be correct, since their totals do not match the row and column marginal frequencies, and therefore, regardless of the actual cell values, it is not possible for the *O* and *E* values to match. There is no basis for assuming that there is an equal number of male and female students within each of the English sections. We have no information that would lead to this expectation, nor is it the hypothesis that we wish to test. The original concern was whether or not the two sections differed in their proportion of male and female students because of possible selective factors. In order to determine specifically if Gwendelyn's class contains a disproportionately large number of male students relative to the college population, we would test just that section with a single-sample χ^2 test (with the *E* values based on the male/female proportion of students eligible to take the course). That, however, would be a separate and different test from the one being considered here.

The χ^2 value is calculated with the same formula used in the single-sample test:

$$\chi^2 = \sum \left[\frac{(O \quad E)^2}{E} \right] \tag{3.3}$$

which leads to the following calculation for Figure 3.2:

$$\chi^2 = \frac{(11-14)^2}{14} + \frac{(13-10)^2}{10} + \frac{(10-7)^2}{7} + \frac{(2-5)^2}{5}$$

$$= \frac{3^2}{14} + \frac{3^2}{10} + \frac{3^2}{7} + \frac{3^2}{5} = 4.6 \tag{3.4}$$

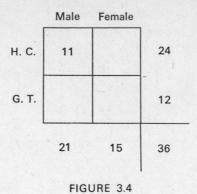

FIGURE 3.4

The likelihood of getting a χ^2 value as large as 4.6 or larger by chance could be determined by entering this value of the statistic in Table F if we had the appropriate degrees of freedom (df). In the single-sample test, df was based on the assumption that the total number of subjects or observations was fixed. In the case of a multi-sample χ^2 test, we assume further that *all* the marginal values, or marginal frequencies, are fixed. In this case, how many of the cells have to be filled in before we can automatically fill in the remaining cells? Look at the situation in Figure 3.4. Marginal frequencies are fixed, and we know how many men are in Harry's class (11); the remaining men $(21 - 11 = 10)$ must be in Gwendelyn's section. Similarly, given that 11 members of Harry's section are men, then the remainder $(24 - 11 = 13)$ must be women. In other words, when all but one item in a row or column have been filled in, the remaining item is fixed. This yields a simple formula for the calculation for the df which is:

$$df = \text{the number of cells providing information} = (C - 1)(R - 1) \quad (3.5)$$

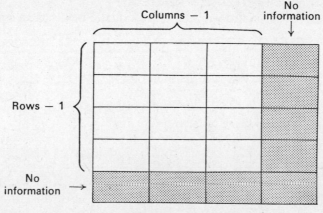

FIGURE 3.5

This formula is illustrated in Figure 3.5. In the example of Harry's and Gwendelyn's sections, we have two rows and two columns; therefore, $df = (2 - 1)(2 - 1) = 1 \times 1 = 1$.

With $\chi^2 = 4.6$ and $df = 1$, the probability value can be determined the same way that it was for the χ^2 single sample. In Table F a χ^2 value of 4.6 with $df = 1$ falls between the table entries for $p = .05$ and $p = .02$. Therefore, the probability of getting a χ^2 value as large or larger than 4.6 by chance is less than .05 (i.e., $p < .05$). How large is this difference in terms of r_m? Table B indicates that for $p < .05$ and a sample size of 36 the value of r_m is between .30 and .35. There is a moderate ($r_m < .35$) and significant ($p < .05$) difference in the proportion of men to women in the two English sections with proportionally more women in Harry's class and more men in Gwendelyn's class.

Can we then go further and attribute these differences to social charms and physical endowments of the two instructors? Before doing so, we would first have to preclude other possible explanations. Among these might be that the time of Harry's section conflicts with certain science or physical education courses taken by many male students. Such a factor provides an explanation on the basis of the hour of the class rather than the nature of the instructor. Another possibility is that Gwendelyn is as notably soft in grading as in other respects, and the male students, who at this particular school are markedly inferior to the females in academic ability, prefer to avoid Harry's more demanding course. The critical factor may, indeed, be the instructor, but not necessarily in the sense that we originally had in mind.

Calculating the value of the statistic and determining the associated p and r_m values is a simple (or at least a simple-minded) matter. It is in the interpretation of the data that sticky problems arise. Given that we find differences between groups that are large and significant, interpreting or making sense of these differences requires additional knowledge of the situation. Statistical manipulations, in themselves, do not tell us what is going on. The formulas and tables deal merely with the numbers fed into them, and, at best, these numbers stand for meaningful aspects of the phenomena being studied. Certainly, the formula cannot "know" about the situation giving rise to these numbers. The statistical calculations, the r_m and the probability values, are *aids* to our interpretation of data. They provide objective measures of the data, which are helpful in describing the basis of an interpretation. The crucial point here is that the individual to whom the meaning of the data is most important, the individual who is most likely to be misled by an improper interpretation, is, of course, oneself. The investigator cannot escape his responsibility to perceive clearly and accurately the situation and make the most rational assessment of the data possible.

The χ^2 test is frequently misused, and the following precautions must be kept in mind:

1. The numbers in the cells refer to category data, that is, the number of

times a particular event has occurred. The numbers cannot be averages or scores. The total number of observations (N) refers to the number of subjects or observations and not to the scores of subjects.

2. The observations are of subjects sampled from a population; each subject must appear only once in the array, or table, so that the value of N refers to the number of *different* subjects in the sample. Measuring each subject twice does not enable us to double the size of N by entering two outcomes for each subject in the table. This restriction is illustrated by the fact that, although we could get a reasonable idea of the weight of the population of students by weighing a sample of 100 students, we would not get the same information by weighing two students 50 times. The use of the χ^2 test in place of the binomial test to check the fairness of a coin does not involve an exception to this rule, since the population we are interested in is that specific coin and its behavior, and the samples are instances of the coin's behavior. No attempt is made to generalize the observations to other coins. This last point is discussed further in supplement F on studies of single subjects.

3. The sum of the expected values in the χ^2 array must equal the sum of the observed values ($\sum E = \sum O$). The precaution here is that both the occurrence and nonoccurrence of the event in which we are interested must be recorded. When tossing a coin, for instance, we base the calculation on both heads and tails rather than just heads or tails.

4. Expected values of less than 5 often lead to an unreliable and inflated value of χ^2. This is most serious when df is equal to 1 and becomes less of a problem as df increases. An extreme example occurs when the expected value is 0, leading to an infinite value of chi square whenever any occurrence of the event is recorded [$(0 - E)^2/E = 1/0 = \infty$].

5. When $df = 1$ (i.e., a 2×2 table or a single-sample test), the correction for continuity (discussed in Chapter 2) should be used. The correction involves the subtraction of 1/2 from the ($O - E$) difference in each cell as given in formula (2.6).

ALTERNATIVE FORMULAS FOR χ^2

There are several algebraically equivalent formulas for calculation of the χ^2 statistic, and these are often more convenient than the definitional formula.

For the 2×2 array, when the cells are labeled in the fashion shown in Figure 3.6, the calculations can be based directly on the cell frequencies:

$$\chi^2 = \frac{N(AD - BC)^2}{(A + B)(C + D)(A + C)(B + D)} \tag{3.6a}$$

or when corrected for continuity,

$$\chi_c^2 = \frac{N(|AD - BC| - N/2)^2}{(A + B)(C + D)(A + C)(B + D)} \qquad (3.6b)$$

A	B	$A + B$
C	D	$C + D$
$A + C$	$B + D$	$N = A + B + C + D$

FIGURE 3.6

The following χ^2 formula for any size array avoids the need for a subtraction for each cell:

$$\chi^2 = \sum \left(\frac{O^2}{E}\right) - N \qquad (3.7)$$

The following formula, which is readily derived from formula (3.7), eliminates the need to calculate expected values. It is based directly on the observed value for each cell (O_{ij}) and the marginal frequencies for the row ($\sum r_i$) and column ($\sum c_j$).

$$\chi^2 = N\left[\sum \frac{(O_{ij})^2}{(\sum r_i)(\sum c_j)} - 1\right] \qquad (3.8)$$

Formulas (3.6), (3.7), and (3.8) are particularly convenient for machine

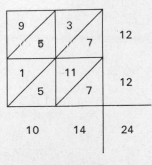

FIGURE 3.7

calculation, and the last is especially useful for either a programming calculator or computer.

To compare the alternative formulas, take the example illustrated in Figure 3.7. By the definitional formula (3.3):

$$\chi^2 = \sum \frac{(O - E)^2}{E} = \frac{(9 - 5)^2}{5} + \frac{(3 - 7)^2}{7} \cdots = \frac{16}{5} + \frac{16}{5} + \frac{16}{7} + \frac{16}{7} = 10.97$$
(3.9a)

and

$$\chi_c^2 = \sum \frac{(|O - E| - 1/2)^2}{E} = \frac{(|9 - 5| - 1/2)^2}{5}$$
$$+ \frac{(|3 - 7| - 1/2)^2}{7} = \frac{12.25}{5} + \frac{12.25}{7} \cdots = 8.40 \qquad (3.9b)$$

By formula (3.6):

$$\chi^2 = \frac{N(AD - BC)^2}{(A + B)(C + D)(A + C)(B + D)} = \frac{24[(9 \times 11) - (3 \times 1)]^2}{(12)(12)(10)(14)} = 10.97$$

and

$$\chi_c^2 = \frac{N(|AD - BC| - N/2)^2}{(A + B)(C + D)(A + C)(B + D)}$$
$$= \frac{24[|(9 \times 11) - (3 \times 1)| - 24/2]^2}{(12)(12)(10)(14)} = 8.40 \qquad (3.10)$$

By formula (3.7):

$$\chi^2 = \sum \left(\frac{O^2}{E}\right) - N = \frac{9^2}{5} + \frac{3^2}{7} + \frac{1^2}{5} + \frac{11^2}{7} - 24 = 10.97 \qquad (3.11)$$

By formula (3.8):

$$\chi^2 = N\left[\sum \frac{(O_{ij})^2}{(\sum r_i)(\sum c_j)} - 1\right]$$
$$= 24\left[\frac{9^2}{(10)(12)} + \frac{3^2}{(14)(12)} + \frac{1^2}{(10)(12)} + \frac{11^2}{(14)(12)} - 1\right] = 10.97 \qquad (3.12)$$

Both formula (3.6), which can be used with a 2×2 table only, and formula (3.8) avoid the need to calculate E values and, therefore, are not subject to rounding errors when the E values are not whole numbers.

NOTE ON THE CORRECTION FOR CONTINUITY

Formula (2.6) gave the correction for continuity for the single-sample test when $df = 1$. It is often suggested that this correction, subtracting 1/2 from the $(O - E)$ difference, be used with formula (3.3) for the 2×2 table situation as well. However, the correction is often too severe for the 2×2 table when E values are very small (i.e., less than 5), making it too difficult to reject the null hypothesis (cf. Grizzle 1967). Somewhat more accurate probability values can be obtained with the Fisher exact test (cf. Siegel 1956). The difference between the "corrected" and uncorrected χ^2 values diminishes as N increases.

TYPE I AND TYPE II ERRORS AND MAGNITUDE OF EFFECT

Again, assume we have a coin and want to decide whether or not it is a fair one. Out of 17 tosses the coin lands heads 13 times. Using the binomial test, we find the probability of the coin being a fair one is equal to .05 and conclude that it is significantly biased in favor of heads. Have we proved that it is actually biased? Clearly not, since there is a finite possibility (just calculated at $p = .05$) that a fair coin would behave in this fashion. Given this situation, we must consider the possibility of its being a mistake to conclude that the coin is biased. In other words, approximately 5 percent of the time the behavior of a fair coin would be such that we would (mistakenly) reject the null hypothesis (H_0) that the coin is fair and conclude that it is indeed biased. Under these circumstances, we would be making an error by rejecting the null hypothesis when, in fact, it is actually true. Such an error is called a **type I error**.

Suppose you have a particular dislike for type I errors. It is a simple matter to reduce the likelihood of this type of error by using a more stringent level of significance, such as .01 or .001. Let us assume that you wish virtually to eliminate the possibility of ever making a type I error in doing research and, therefore, decide not to reject the null hypothesis unless the obtained probability value is less than one in a million. Working under these conditions, you spend your entire scientific career (which turns out to be a remarkably short one) without ever getting data that justify rejecting the null hypothesis. You encounter many interesting phenomena, but none of them ever gives a probability value of less than .000001. You decide something has gone wrong. Certainly it cannot be that you have made type I errors, so you must be making some other kind of error—type II errors. A **type II error** is one in which you *fail* to reject the null hypothesis when it is false. You

Result of the Statistical Test

		H_0 Not Rejected	H_0 Rejected
The "Real" Situation	H_0 True	OK	Type I Error
	H_0 False	Type II Error	OK

FIGURE 3.8

fail to detect differences or effects that are actually present. The relationship between the decision concerning H_0 and type I and type II errors is summarized in Figure 3.8.

Any attempt to reduce type I errors by using a more stringent level of significance raises (to a generally unknown or unspecifiable degree) the likelihood of making type II errors. The reverse is also true; any attempt to reduce type II errors by using a more lenient level of significance increases the likelihood of making type I errors. Fortunately, the ability to detect differences or effects while keeping both type I and type II errors at an acceptable level increases as the sample size, or number of observations, increases. Table B shows that even small differences (in terms of r_m) can be detected as being significant when N is large. Improving the experimental design and controls also serves to reduce both type I and type II errors by reducing the influence of chance factors.

The balance between type I and type II errors for a given situation is always a function of the relative cost (in terms of how much we can stand to lose) of making these errors. For example, in searching for a drug to cure an intractable form of cancer, a relatively lenient significance level, perhaps .10 or even .20, would be appropriate. Under these conditions we might frequently commit type I errors and mistakenly conclude that a particular drug is useful. This would lead to testing the drug further on additional patients. We would then discover our error, but the cost of the testing would not be very great. On the other hand, if a stringent level of significance (i.e., .01 or .001) were used, we could very easily ignore a drug that is somewhat effective and with further research and refinement might well be very useful. The cost of making a type II error in this instance would be very high in terms of the lives that could be saved if the effective drug were available. A different situation holds for ESP research. A reliable and valid demonstration of telepathy, precognition, clairvoyance, or psychokinesis would result in a marked revision not only of psychological laws but also of the physical laws of the universe. Therefore, it is appropriate to use a highly stringent level of significance in order to avoid type I errors (which would lead to a major revision of scientific laws), while permitting a fairly large chance for type II errors (which would delay the acceptance of ESP phenomena).

The consideration of type I and type II errors is, of course, only one aspect of the interpretation of data. In any situation in which there are differences between the groups, any desired probability value can be obtained by testing an appropriately large number of subjects. Therefore, our decision to act *as if* the effect is "real" must be the function of the magnitude of effect (r_m) as well as of the probability value. The term "real" in this context means that the effect is sufficiently large (in terms of r_m and p) for us to talk about our research or plan future research on the basis of the effect. Typically, when we give a formal report of research (i.e., publish a paper), we use stringent criteria, since the effect must be clear and unambiguous enough to convince the skeptical or even the hostile reader.

Let us, then, reconsider type I and type II errors substituting the term "low r_m value" for the term "lenient level of significance" and "high r_m" for "stringent." In the cancer research example, we would be interested in any drug that appears to be reasonably effective (has a high r_m value) even if the effect is not quite significant. On the other hand, in the typical ESP experiment, which requires many trials, usually hundreds or thousands, to detect the phenomenon, the r_m value is very low ($< .10$) though the probability value may also be very striking ($p < .001$). Once again, the final responsibility for the interpretation of the data and the situation rests with the experimenter. It cannot be transferred to the numbers or numerical calculations. The interpretation of data with the aid of p and r_m values is discussed in greater detail in Chapter 12.

Somewhat tangential to the preceding discussion is an interesting situation in which there is general agreement that we can tolerate *no* type I error and therefore are willing to accept an appreciable type II error. This unique circumstance occurs in the original Anglo-Saxon jury trial. Since the jury must agree unanimously in order to give a verdict of guilty, a reasonable doubt by any one of the jurors is enough to prevent conviction. The rationale of this requirement can be seen in the following terms. To obtain the sum of 12 numbers, we would add them up. To check the accuracy of that sum, we would add the column again and see if the two sums match. If they do, we assume the sum is correct. We make the reasonable assumption that it is unlikely for random errors to have led to the same result both times. In other words, we assume that, when dealing with an imperfect process, we can get agreement only on the *truth*. Note that the term Truth (with a capital T) is the subject matter of theologians or philosophers and not of scientists, who, at best, deal with a more limited and fragile empirical truth. In the jury trial, we assume that the only thing that 12 independent, impartial men can agree upon would be the truth. Although the system in practice occasionally fails and permits a type I error—convicts an innocent person—no satisfactory substitute has yet been proposed. With such a jury system we clearly run a very high risk of many type II errors—failing to convict individuals who are really guilty. A democratic society is necessarily willing to accept this penalty to avoid placing the vast majority of the citizens in terror of the law and the courts. Simple proposals to facilitate convictions by easing the requirements of the jury system are equivalent to using a more lenient level of significance and would invariably lead to a greater likelihood of type I errors and convictions of innocent individuals. By analogy to the use of improved experiments, an appropriate method to reduce type I errors would be more efficient and scientific collection of evidence coupled with improved legal procedures.

SIGNIFICANCE IN THEORY AND PRACTICE

The term "level of significance" actually has two meanings. As it has already been defined, a significant effect is one that is sufficiently unlikely to have occurred by chance for us to be willing to *assume* that it is a non-chance phenomenon. In other words, if we find the results convincing, then the results are "significant" in the sense defined. Clearly, a measure of the magnitude of effect (r_m) aids in deciding whether or not the observed difference is trivial. Depending on the situation and the sample size, a probability value of .10 might be significant in one case, and a value of $p = .01$ might be required in another case.

When you report results of a study, it is important, regardless of the level of significance employed, that your audience be informed of the size of the difference as well as the probability and magnitude of effect values, so that the members of the audience can make their own decisions concerning the data. The term "significant" is more generally (if inaccurately) used to indicate the probability value that you think the audience will accept as significant. The audience may be individuals to whom you describe the results or, more importantly, the editors of the journals to which you submit the experimental reports. For this second meaning of significant, a probability level of 5 percent $(p = .05)$ is generally suitable regardless of the particular situation or the sample size. In practice, it is often necessary to make a study suitable for public display by adding subjects, for a given experimental effect, so as to reduce the probability value to .05. There is definitely a gamelike quality to this second meaning of significance.

† NORMAL CURVE APPROXIMATION FOR χ^2 WHEN *df* > 30

Table F can be used to determine the significance of an obtained χ^2 value if *df* is not greater than 30. With a 7 × 7 table (*df* = 6 × 6 = 36), some other means of determining the probability value for χ^2 is needed. Fortunately, when *df* increases, the sampling distribution for χ^2 approaches a normal distribution, and the relationship between Z and χ^2 can be approximated by:

$$Z = \sqrt{2\chi^2} - \sqrt{2\,df - 1} \qquad (3.13)$$

For example, with $df = 36$, $\chi^2 = 60$, and $N = 300$:

$$Z = \sqrt{2(60)} - \sqrt{2(36) - 1}$$

$$= \sqrt{120} - \sqrt{71} = 2.53 \qquad p < .02 \qquad\qquad (3.14)$$

χ^2 MULTI-SAMPLE

Formula

a. $\chi^2 = \sum \left[\dfrac{(O_{ij} - E_{ij})^2}{E_{ij}} \right]$

where $E_{ij} = \dfrac{(\sum r_i)(\sum c_j)}{N}$

$df = (\text{rows} - 1)(\text{columns} - 1)$

b. $\chi^2 = \sum \left[\dfrac{(O_{ij})^2}{E_{ij}} \right] - N$

(no subtraction for each cell)

c. $\chi^2 = N \left[\sum \dfrac{(O_{ij})^2}{(\sum r_i)(\sum c_j)} - 1 \right]$

(no separate calculation of E values)

d. When $df = 1$, use the correction for continuity:

$\chi_c{}^2 = \sum \dfrac{(|O_{ij} - E_{ij}| - 1/2)^2}{E_{ij}}$

e. When $df = 1$ and the cells are labeled

A	B	$A + B$
C	D	$C + D$
$A + C$	$B + D$	

$\chi_c{}^2 = \dfrac{N(|AD - BC| - N/2)^2}{(A + B)(C + D)(A + C)(B + D)}$

Table F   provides p values based on χ^2 and df.

Table B   provides r_m values based on p and number of cases (N).

Table C   provides r_m values based on χ^2 and N.

Restrictions   All observations should be independent with $N = \sum O =$ number of different subjects tested. Individual subjects cannot be tested more than once.

The expected (E) values for all cells should be at least 5 to avoid inflated χ^2 values. No cell can have an E value of 0.

When $df = 1$, the correction for continuity should be used. The χ_c^2 value leads to more accurate p values than χ^2 but underestimates the magnitude of effect (r_m). When precise r_m values are required, use χ^2 in place of χ_c^2 in Table C or use the r_ϕ coefficient (Chapter 12) in place of r_m.

Procedure

1. Obtain E values for each cell by multiplying the marginal frequencies of the row $(\sum r_i)$ and column $(\sum c_j)$ containing that cell and dividing the product by N. $E = (\sum r_i \sum c_j)/N$.
   Note: Calculation of E values is not necessary when using formula (c) or (e).
2. For each cell determine the term $(O_{ij} - E_{ij})$, square the term, and divide by the expected (E) value for the cell.
   Note: With formula (c), $(O_{ij})^2$ is divided by the product of the marginal frequencies for each cell.
3. Add the value obtained in step 2 for all cells and complete the calculation of χ^2.
4. Use Table F to obtain p values based on χ^2 and df.
5. Obtain r_m values from Table B based on p and N or Table C based on χ^2 and N.

Example 1

$df > 1$

The majors of college men and women with equivalent ability, in terms of IQ and College Board scores, are:

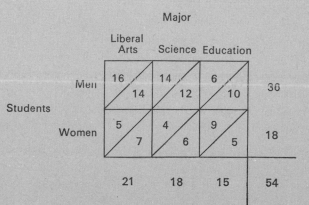

Major

		Liberal Arts	Science	Education	
	Men	16 14	14 12	6 10	36
Students	Women	5 7	4 6	9 5	18
		21	18	15	54

Formula

a. $\chi^2 = \sum \left[\dfrac{(O - E)^2}{E} \right]$

$\quad = \dfrac{(16 - 14)^2}{14} + \dfrac{(14 - 12)^2}{12} \cdots$

$\quad = \dfrac{2^2}{14} + \dfrac{2^2}{12} + \dfrac{4^2}{10} \cdots = 6.66$

Formula

b. $\chi^2 = \sum \left[\dfrac{(O_{ij})^2}{E_{ij}} \right] - N = \dfrac{16^2}{14} + \dfrac{14^2}{12} \cdots - 54$

$\quad = 60.6571 - 54 = 6.66$

Formula

c. $\chi^2 = N \left[\sum \dfrac{(O_{ij})^2}{(\sum r_i)(\sum c_j)} - 1 \right]$

$\quad = 54 \left[\dfrac{16^2}{(21)(36)} + \dfrac{14^2}{(18)(36)} + \dfrac{6^2}{(15)(36)} \cdots -1 \right]$

$\quad = 54(1.123280 - 1) = 6.66$

From Table F, $p < .05$ with $df = 2$.
From Table B, $r_m > .25$ with $p < .05$ and $N = 54$.
From Table C, $r_m > .30$ with $\chi^2 = 6.66$ and $N = 54$.

Conclusion

There are significant ($p < .05$) but moderately small ($r_m > .30$) differences in the proportions of men and women students majoring in liberal arts, science, and education. The men in the sample tend to prefer liberal arts and science, and the women tend to prefer education.

Example 2

$df = 1$
One statistics course is taught with teaching machines, and an equivalent course is taught with a standard text and lectures. The number of students that pass and fail a common final examination is measured:

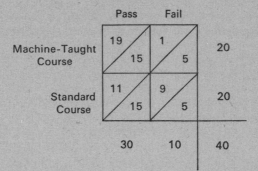

Formula

d. $\chi_c{}^2 = \sum \dfrac{(|O_{ij} - E_{ij}| - 1/2)^2}{E_{ij}}$

$= \dfrac{(|19 - 15| - 1/2)^2}{15} + \dfrac{(|1 - 5| - 1/2)^2}{5} + \cdots$

$= \dfrac{(3.5)^2}{15} + \dfrac{(3.5)^2}{5} + \cdots = 6.53$

Formula

e. $\chi_c{}^2 = \dfrac{N(|AD - BC| - N/2)^2}{(A + B)(C + D)(A + C)(B + D)}$

$= \dfrac{40(|(19)(9) - (1)(11)| - 40/2)^2}{(20)(20)(30)(10)}$

$= \dfrac{40(160 - 20)^2}{120,000} = \dfrac{40(140)^2}{120,000} = 6.53$

From Table F, $p < .05$ based on $\chi^2 = 6.53$ and $df = 1$.
From Table B, $r_m > .30$ based on $p < .05$ and $N = 40$.
From Table C, $r_m > .35$ based on $\chi^2 = 6.53$ and $N = 40$.

Conclusion

There is a moderate ($r_m > .35$) and significant ($p < .05$) difference in the proportion of students passing in the machine-taught class as compared to the standard course. The machine-taught class had a higher proportion of passing students.

SUMMARY

1. *Multi-sample comparisons* involve testing two or more samples to determine if they can reasonably be assumed to be random samples from the same population. In contrast to single-sample tests, no prior knowledge of the characteristics of the population is needed.

2. The *correction for continuity* is only used when $df = 1$ (i.e., for 2×2 tables and for two-cell single-sample tests). When sample sizes are small, the correction may be too severe (making it too difficult to reject H_0). The Fisher exact test (cf. Siegel 1956) can be used in these situations.

3. *Type I errors* are made when H_0 is mistakenly rejected. *Type II errors* occur when H_0 is not rejected even though it is false. The probability of making type I errors is equal to the level of significance, but using a more stringent level of significance will lead to an increase in type II errors.

4. A statistical measure is said to be *significant* if the probability of its occurring by chance is so small that we can assume it is a non-chance effect, or that H_0 is not correct. In practice, a .05 or .01 level of significance is usually used. The *interpretation* of an outcome is a function of both the probability (p) and size of the effect (r_m).

† Probability values for χ^2 values are given in Table F for cases when df is not greater than 30. A *normal curve approximation* can be used when $df > 30$.

chapter four

CATEGORICAL DATA, SCORES, RANKING, AND ORDINAL DATA

The previous chapters on the χ^2 and binomial tests dealt with measurements of the number of individuals or observations that could be placed into certain categories. Measurements based on categorizing, or naming of the subject are called **categorical** data. Data or observations may be categorized on the basis of sex (male/female), status (freshman/sophomore/junior/senior), or performance (pass/fail). In many cases categorical data are a very rough form of measurement. For example, if pass/fail categories are used for exam results, certain students in the pass category may have near perfect grades, while others may be barely above the cutoff point for passing. It is often feasible to refine such crude measurements to provide more information. For instance, if the number of categories is increased from pass/fail to include high pass/medium pass/low pass/high fail/medium fail/low fail, it is possible to distinguish among students much more precisely. Each of these categories can be subdivided further to provide even finer distinctions. Finally, if we make each category as small as the finest division used to grade the exams, we will end up with 100 categories, which might be conveniently labeled as $0/1/2/3/\cdots 98/99/100$. In other words, we have gone from category measurements to numerical values, or **scores**. More information is available from scores, which represent fine distinctions, than from categorical data, which are based on relatively few categories.

The next problem is to generate the appropriate statistical tests to analyze score data. The simplest way of dealing with scores is to *order* them in terms of their relative size (usually from lowest to highest) and to base the statistical analysis upon this ordering, or **ranking**, of the scores. The statistical tests available to handle ranked, or **ordinal**, data are quite efficient, easy to use, and have the additional advantage of forming a very good foundation or introduction to the statistical procedures used in dealing with the scores as numbers (Chapter 7).

COMPARISON OF TWO INDEPENDENT GROUPS

Assume we wish to test two independent groups of subjects (i.e., subjects randomly selected from a population and randomly assigned to groups) on a series of simple problems. One group, called the control group, is told only that we want to get an idea of how well the average student can solve a given set of problems. The other group, the experimental group, is told that the problems are from a new test, which has been useful in predicting intelligence, moral fitness, and sexual potential. We also casually mention that the Dean of Students has expressed interest in the test and has requested a copy of all the test scores. We *assume* that the experimental group is going to have more anxiety when taking the test than the control group.

RANDOM SAMPLING

A necessary precondition for this study is that the two groups be equivalent *initially*, the only differences between them being those imposed by the conditions of the experiment. If this is the case, then any differences between the final results of the two groups can presumably be attributed to the effects of the experimental manipulation (the instructions). These two groups should also be representative samples from the same population, that is, the population of all students. One way of making the two samples comparable, or equivalent, would be to adjust the membership of each sample so that the overall characteristics of the two are the same. This is difficult to accomplish and not commonly used. The easiest way to obtain two or more reasonably equivalent and representative samples is to use *random sampling*. Subjects may be selected on some chance basis (such as picking names out of a hat) or on some basis unrelated to the relevant aspects of the study (such as selecting every third name in an alphabetical list). These procedures certainly do not ensure that the two samples will be identical, but they do ensure that they will differ on a chance, or random, basis only. Under these conditions factors that might affect the outcome of the study, such as the students' problem-solving ability or general intelligence or motivation, would tend to be approximately equal for both groups.

THE MEDIAN AND THE MEDIAN TEST

The number of problems correctly solved by the members of the control and experimental groups are shown in Table 4.1 where the scores of the two groups are also ranked in order of size. This ranking, or ordering, which will be discussed in the next section, makes it very easy to see that the scores of the experimental group tend to be higher overall than the scores of the control group. We must now decide whether this difference is sufficiently large to assume that it is unlikely to be merely a chance difference—a difference that would be expected from randomly drawn samples from the same population.

We already know how to deal adequately, if inefficiently, with this situation. Assume that the two groups are indeed random samples from the same population. Then both high and low scores would be equally likely to appear in the two groups. If we were to separate the higher half and the lower half of the entire group of scores (i.e., all 20 scores), the dividing line should split *both* samples in half. The dividing line that splits the *entire* group of scores into two equal portions is called the **median**, or middle score. In the example the median lies halfway between the tenth and eleventh scores and is indicated by the line drawn in Table 4.1. The analysis up to this point is summarized

TABLE 4.1

Control Group			Experimental Group	
Score	Rank		Score	Rank
2	1			
3	2			
5	3			
7	4.5			
7	4.5			
8	6			
9	7			
11	8			
13	9			
			14	10
—	←——— Median ———→		—	
15	12		15	12
			15	12
			18	14
			19	15
			20	16.5
			20	16.5
			22	18
			24	19
			25	20

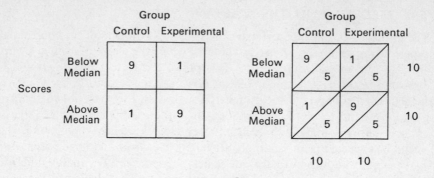

FIGURE 4.1

FIGURE 4.1

in Figure 4.1, and the data are now suited to a χ_c^2 statistic, which can be calculated as follows:

$$\chi_c^2 = \frac{(|9-5|-1/2)^2}{5} + \cdots = \frac{49}{5} = 9.8 \qquad (4.1)$$

$$p < .01 \qquad r_m > .54$$

We have effectively reduced the data from ordinal data, or scores, to categorical data (the categories being above and below the common median), and the χ^2 test is appropriate. Since the calculated χ_c^2 value of 9.8 (with $df = 1$) is significant at the .01 level, and $r_m > .54$, we conclude that there is a significant difference between the properties of scores for the control and experimental groups that fall above and below the common median; the subjects in the experimental group tend to have the higher scores and the subjects in the control group tend to have the lower scores. In other words, the scores of the experimental group are significantly and appreciably higher than those of the control group. This test, called the **median test**, can, therefore, be used to analyze ranked data from two independent groups by converting the scores to categorical data.

Consider a similar experiment for which the scores are given in Table 4.2 and the median test data are shown in Figure 4.2. The value for χ_c^2 is:

$$\chi_c^2 = \frac{(|7-5|-1/2)^2}{5} + \cdots + \cdots = \frac{9}{5} = 1.80 \qquad (4.2)$$

$$p > .10 \qquad r_m < .36$$

In this case, the median test gives a nonsignificant and low χ_c^2 value even though the scores for the experimental group appear to be higher than those for the control group. The problem with the median test is that, when the ordinal data are changed to categorical data, any differences among the scores above and below the median are not considered, and this results in the loss of valuable information. Thus we need a procedure that utilizes more of the information available in the ranked scores. The median test is not, in

TABLE 4.2

Control Group		Experimental Group	
Score	Rank	Score	Rank
2	1		
3	2		
6	3		
7	4.5		
7	4.5		
8	6.5		
8	6.5		
		9	8
		10	9.5
		10	9.5
11	11		
13	12		
15	13		
	$\Sigma = 64$	17	14
		18	15
		20	17
		20	17
		20	17
		21	19
		25	20
			$\Sigma = 146$

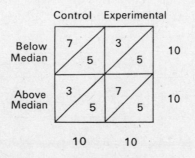

FIGURE 4.2

general, the preferred test for this type of example, and it was used here only to introduce the following test.

THE WHITE TEST, OR MANN-WHITNEY U TEST

A clearer picture of the relative positions of the two sets of scores in Table 4.2 can be obtained by dealing more directly with the ranks of the scores. The lowest score is given the rank of 1, and the highest and last score is given a rank of 20. When two scores have the same value (such as the two 7s in the control group), we cannot distinguish between the two by giving one a rank of 4 and the other a rank of 5. Therefore, we give each score the average of the two ranks $[(4 + 5)/2 = 4.5]$; the sum of the ranks of the two scores $(4.5 + 4.5 = 9)$ is the same as it would be if we could give the two scores different ranks $(4 + 5 = 9)$. Similarly, when there are three identical scores (which can be in the same or different groups), they are given the average of the three ranks that would be assigned to them if the three scores could be distinguished. For example, the three scores of 20 in the experimental group all have the rank of 17.

Given these data, what would we expect if the null hypothesis—that differences in the two sets of scores are due to chance factors only—were true? There would be no reason to expect the high scores to concentrate in one or the other group. In fact, the most probable outcome would be for each group to have both high- and low-ranked scores with the scores of the two groups well intermingled. Therefore, we would expect the sum of the ranks of the scores for each of the two groups to be essentially equal, or if the number of subjects in the groups is not equal, we would expect the average rank for the two groups to be equal. In other words, if the null hypothesis were correct, there should be an appreciable overlap of scores from the two groups, and the high scores should not concentrate in one group with the low scores in the other. A simple but suitable measure, or statistic, for this analysis is the smaller of the two sums of ranks for the two groups. For the data in Table 4.2 how often would we expect to get an outcome in which the two groups are so different as to have rank sums of 64 and 146 when, by chance, we would expect the sums for each group to be approximately 105, $[(64 + 146)/2 =$

TABLE 4.3

Maximum Value of T for $p = .05$

$n_2 \downarrow$ $n_1 \rightarrow$	2	3	4	5	⑥	7	8	9	⑩	11	12	13	14	15
4			10											
5		6	11	17										
⑥			7	12	18	㉖								
7			7	13	20	27	36							
8		3	8	14	21	29	38	49						
9		3	8	15	22	31	40	51	63					
⑩		3	9	15	23	32	42	53	65	㉘				
11		4	9	16	24	34	44	55	68	81	96			
12		4	10	17	26	35	46	58	71	85	99	115		

26 is maximum value of T for $n_1 = 6, n_2 = 6$ · · · 78 is maximum value of T for $n_1 = 10, n_2 = 10$

210/2 = 105]? Is it likely that two random samples from the same population would differ to this degree?

As with χ^2 we need a sampling distribution of the statistic in order to answer these questions. This has been calculated, in the form of the **U test** by Mann and Whitney (1947) and in an algebraically equivalent but simplified form by White (1952). Table I gives the probability values associated with the smaller of the two sums of ranks (T) for two samples having a total of 30 or fewer subjects. A simple approximation formula permits dealing with larger sample sizes. The relevant section of the table is shown below in Table 4.3. For example the table entry indicates that, when the size of each of the two samples, or groups, is 6 ($n_1 = 6$ and $n_2 = 6$), a smaller rank sum of 26 or less would be significant at the 5 percent level. In other words, if the sum of the ranks of one group was 26 (and 52 for the other), we would conclude that this difference would not have occurred by chance more often than 5 percent of the time. The data in Table 4.2 gave values of 64 and 146 for the rank sums of the two groups; 64 is smaller than the maximum T value (78) for $p < .05$ and is almost small enough to be significant at $p < .01$. The difference between the two groups is significant at the 5 percent level, and $r_m > .44$ (based on $20 - 2 = 18$ as the sample size). Therefore, there is a fairly large and significant difference between the scores of the two groups, with the experimental group tending to get higher scores than the control group. The sample size measure for the estimation of r_m is reduced by 1 for each group; as we will see in Chapter 12, the measure of sample size is based on degrees of freedom (df).

This covers the essential features of the White test, but consider the problem shown in Table 4.4. The scores are ranked, as in the previous example, from lowest to highest score, giving a rank sum of 16 for the control group and 29 for the experimental group. Notice that, since the group

TABLE 4.4

Control Group	Experimental Group
	9
	8
	7
6	
	5
4	$\Sigma = 29$
3	
2	
1	
$\Sigma = 16$	

with the fewest subjects tends to have the highest rankings, the difference between the two sums is small, which is misleading. Although we could use the average ranks of the two groups (by dividing the sum of ranks by the number of scores in the group), a simpler procedure is to arrange the rankings so that the group with the fewest subjects tends to get the smaller rankings. In the present example we can reverse the rankings in the fashion given in Table 4.5, by starting the highest score with rank 1 so that the smaller ranks tend to fall within the group with the fewest subjects, the experimental group. The difference between 11 and 34 is greater than between the 16 and 29 obtained before.

TABLE 4.5

Control Group	Experimental Group
	1
	2
	3
4	
	5
6	$\Sigma = 11$
7	
8	
9	
$\Sigma = 34$	

Obviously, the previous statement regarding the statistic for the White test should be revised. Instead of being merely the smaller of the two sums of ranks for the two groups, the statistic is the sum of ranks for the group with the fewest subjects when the scores are ranked to give this group the smaller of the two possible values. In other words, we can rank either from the smallest to the largest or from the largest to the smallest scores, and we want to use the ranking that will provide the smallest sum possible for the group with the fewest subjects. In the present example the value of the statistic is 11, *not* 29 (which is the sum of ranks for the group with fewest subjects when they are ranked from lowest to highest), nor is it 16 (the sum of ranks for the group with the most subjects). In practice, we do not have to rank both ways, since all the information necessary is contained in the first ranking. If the sum of ranks for the smaller group (T) is known, then we can calculate the sum for the reversed rankings by using formula (4.3):

$$T^1 = [n_1(N + 1)] - T \qquad\qquad (4.3)$$

where n_1 = the size of the smaller group
        N = the total number of subjects in both groups = $n_1 + n_2$

In our example $T = 29$. Thus we obtain:

$$T^1 = [4(9 + 1)] - 29 = 40 - 29 = 11 \tag{4.4}$$

which is the same value (11) that we calculated previously. Remember when using Table I that n_1 is always the group with the fewest subjects; n_2 is the other group; and the value to enter in the table is T or T^1, whichever is smaller. When the two groups are of equal size ($n_1 = n_2$), T^1 for a given value of T always works out to be the sum of ranks of the other column. For example, in the problem shown in Table 4.2, we would have:

$$T^1 = [10(20 + 1)] - 64 = 210 - 64 = 146 \tag{4.5}$$

which is as it should be.

To return to the example of the two groups of subjects who have to solve a series of problems, if the two groups were comparable before the experiment, then the observed difference is presumably due to the experimental manipulations. However, although we have assumed that the experimental group is more anxious than the control group, we cannot be sure without further testing. In fact, even if the difference is due to the manipulations, we have no indication of what aspect of the manipulations led to the difference. It may well be that the subjects in the experimental group hardly believed a word of our somewhat heavy-handed description of the test and were, therefore, generally suspicious of the experiment and felt that they better do their best. One advantage of working with human subjects (as opposed to lower animals) is that, after the data are collected, we can ask the subject to give his thoughts about the experiment. This information must be interpreted cautiously but is often valuable in detecting what aspect of a situation the subject really responded to.

† THE WHITE TEST WHEN $n_1 + n_2 > 30$

When $n_1 + n_2 > 30$, Table I cannot be used to determine probability values. The sampling distribution approaches the normal as $n_1 + n_2$ increases, and the critical points (i.e., maximum value of T to be significant) can be obtained from formula (4.6):

$$T = \left[\frac{n_1(n_1 + n_2 + 1)}{2} \right] - \left[Z \sqrt{\frac{n_1 n_2 (n_1 + n_2 + 1)}{12}} \right] \tag{4.6}$$

where $Z = 1.96$ for the 5 percent level and
             2.58 for the 1 percent level

RELATED-SAMPLES TESTS

Testing independent groups is not the only way to compare two experimental conditions. There may be advantages to testing individual subjects under each of the two conditions. Or we may want to pair subjects by matching individuals that are essentially equivalent and then test the members of each pair under different experimental conditions. We would use these procedures when we are concerned with the change in an individual's score from one condition to another or with the differences between the scores of matched individuals. Some advantages and problems of using a related-samples experimental design will be discussed at the end of this chapter.

SIGN TEST

Suppose we would like to compare the effects of noisy or quiet working conditions on the ability of individuals to solve simple math problems. Testing each of 10 subjects under both quiet and noisy conditions provides the set of scores given in Table 4.6. It is clear that most subjects solve more

TABLE 4.6

Subject	Score under Quiet Conditions	Score under Noisy Conditions	Change
a	20	16	−
b	18	15	−
c	18	13	−
d	16	12	−
e	14	12	−
f	13	14	+
g	12	9	−
h	10	7	−
i	8	8	0
j	7	3	−

problems under the quiet condition than under the noisy condition, but is it likely that this difference occurred by chance? A very simple way of dealing with this problem is to consider only the direction of the change of each subject's score in going from one condition to the other. In the data 8 out of 9 subjects showed a lower score under the noisy condition than under the quiet condition. The tenth subject had the same score under both conditions, and we arbitrarily throw this score out, since it fits in neither the increase nor the decrease category. If the null hypothesis were true, if noise level had no effect upon the performance, we would not expect the subjects to tend to score higher under one condition than the other. Although a given subject would not necessarily score the same under both conditions, or even under the same condition, since there may be random or extraneous factors operating, he would be as likely to score higher under the noisy condition as under the quiet condition—if the noise level made no difference. Therefore, we would expect half the subjects to have higher scores under the quiet condition; that is, the probability of a subject getting a higher score under the quiet condition is equal to 1/2. When stated in these terms, the problem of analysis is changed from one dealing with scores to one dealing with categorical data (increase or decrease). This conversion from scores to categorical data by means of the **sign test** is analogous to the use of the median test for two independent groups, except that now the appropriate categorical-data test is the binomial test. The binomial test indicates that $p < .05$ for 8 out of 9 changes in the same direction—this is equivalent to tossing a coin 9 times and getting 8 heads. The r_m value for this outcome is .60. There are large and significant differences in performance of the subjects under the quiet compared to the noisy condition, with the higher scores tending to occur under the quiet condition.

MATCHED-PAIRS SIGNED-RANKS, OR WILCOXON, TEST

In the previous section, we found that in the reduction of ordinal data to categorical by the median test some information was neglected, which resulted in a relatively weak or inefficient test. The same holds true for the sign test, as shown in Table 4.7. Here again it looks as if these scores tend to be higher under the quiet than noisy condition, but the difference is not as striking as in the previous example. With 8 out of 10 subjects showing higher

TABLE 4.7

Subject	Score under Quiet Conditions	Score under Noisy Conditions	Change
a	25	15	−
b	24	8	−
c	22	10	−
d	19	13	−
e	19	4	−
f	16	19	+
g	15	2	−
h	13	14	+
i	12	4	−
j	10	5	−

scores under the quiet condition, the sign test gives a $p > .10$. However, the *sign* of the difference (increase or decrease) does not take into account the *size* of the difference between a subject's two scores. If we look at that information, given in Table 4.8, the picture becomes more interesting. Although 2 of the subjects did show higher scores under noisy than quiet conditions, these differences were the two smallest differences obtained in the sample.

TABLE 4.8

Subject	Score under Quiet Conditions	Score under Noisy Conditions	Difference $(N - Q)$	
a	25	15	-10	
b	24	10	-14	
c	22	12	-10	
d	19	13	$- 6$	
e	19	6	-13	
f	16	19		$+3$
g	15	7	$- 8$	
h	13	14		$+1$
i	12	4	$- 8$	
j	10	5	$- 5$	

What we need is a test that will take into account not only the direction of the difference between the scores of a subject, but also the size of this difference. A simple procedure would be to rank each of the differences (paying no attention to the sign of the difference) and then sum the ranks for the plus differences and for the minus differences. If the null hypothesis were true, that the two experimental conditions are essentially the same, then, as in the sign test, we would expect the scores for the subjects to be basically the same under the two conditions. Since plus and minus changes would be equally likely to occur and the average size of the changes in each direction would be equal, we would expect the sum of the ranks of the plus changes to be equal to the sum of the ranks of the minus changes.

The calculation for the example data of Table 4.8 is shown in Table 4.9. The sum of the ranks for the minus changes is 52, and the sum of the ranks for the positive changes is equal to 3. Under the null hypothesis we would expect the sum of the ranks for the changes in both directions to be equal to 55/2 or 27.5. How likely is it that we would get values as unequal as 52 and 3 when we expected 27.5 and 27.5? For convenience, we use the smaller of the two sums as our statistic, in this case 3.

As with other tests, a sampling distribution for this statistic could be

TABLE 4.9

| Subject | Difference ($N - Q$) | Rank | |
		Minus Changes	Plus Changes
a	−10	7.5	
b	−14	10	
c	−10	7.5	
d	−6	4	
e	−13	9	
f	+3		2
g	−8	5.5	
h	+1		1
i	−8	5.5	
j	−5	3	
	sum of ranks	52.0	3

generated—in this case by randomly selecting pairs of scores, determining the difference between the scores of each pair, ranking the differences without reference to sign, and finally summing the ranks of the plus differences and the minus differences. By repeating this process with sets of 10 pairs each, we would eventually obtain a sampling distribution depicting the chance outcomes. In general, most of the values for this statistic (called T) of the **matched-pairs signed-ranks test** would be reasonably close to 27.5. The more deviant (i.e., the lower) the value is, the less often it would occur by chance.

The sampling distribution for the T statistic has been calculated by Wilcoxon (1949) and is shown in Table J. We can see that with 10 subjects a T value no larger than 8 would be significant at the 5 percent level and no larger than 3 would be significant at the 1 percent level. Since the obtained value of 3 is, therefore, likely to occur by chance no more than 1 percent of the time, we would reject the null hypothesis. The r_m value for $p = .01$ and 10 subjects is .73. Note that the measure of sample size, used in Table B to obtain r_m for this test, is equal to the number of pairs minus one—as with the White test this reflects the use of the df. Therefore, there is a large ($r_m = .73$) and significant ($p < .01$) difference in performance under the two conditions, with better performance being obtained under the quiet condition. The matched-pairs signed-ranks test, or **Wilcoxon test**, is an efficient and easily used test. However, related-samples tests involve certain problems which should be discussed briefly at this point.

REPEATED-MEASURES AND MATCHED-GROUPS DESIGNS

The most common experimental designs employ independent groups of subjects. Under certain circumstances, however, it is preferable to use the same subject under two or more conditions (**repeated-measures design**) or to use subjects who are first matched in pairs or groups, with one subject of a group being tested under each condition (**matched-groups design**). Subjects are generally used more than once in an experiment when they are few in number (e.g., individuals who have certain types of brain damage or who are left-handed albinos with genius IQ), when they are expensive to obtain (e.g., Panda bears), or when they are expensive or difficult to produce (e.g., subjects requiring appreciable prior training or surgery). Under these conditions the relatively large number of subjects needed for studies of independent groups are not available, and it is necessary to obtain all the information possible from the few subjects at hand.*

In a repeated-measures design, however, there is always the possibility of sequential effects—factors that cause the subject to change as a result of his being tested. Warm-up, practice, boredom, or fatigue are effects that tend to cause performance on the second test to differ from the first, independent of the experimental manipulation. If these sequential effects are relatively large, they may obscure the effects of the experimental variable. Often they can be reduced by varying the order of presentation of the experimental situations from subject to subject. In the example of the noisy and quiet working conditions, for instance, half the subjects would be tested under the quiet condition first and then the noisy, and vice versa for the other subjects. This procedure serves to balance out, or *counterbalance*, the sequential effects. Warm-up sessions, practice, and rest periods can also be used to minimize the sequential effects in an experiment. In drug studies, sufficient time must be allowed for aftereffects to wear off before the subject is retested.

A common type of repeated-measure design involves testing each subject before and after a specific variable is introduced. But if there is an appreciable interval between each measurement, differences between the scores may not necessarily be attributable to the intervening experimental variable. For example, would it be reasonable to conclude that brand X aspirin is effective in reducing headaches on the basis of data showing that 80 percent of subjects with headaches reported their head pain had gone within 24 hours of receiving the aspirin? Such a conclusion is clearly suspect, particularly if we recall that one of the most effective weapons in the bristling arsenal of modern medicine is known (informally) as "tincture of time." In general, in a before-after design, great care has to be taken that the observed outcome is not the result of some time-dependent process independent of the experimental manipulation.

The usual way of dealing with known and unknown extraneous factors is to use a reference, or control, group in a matched-samples or independent-groups design. Subtle phenomena seen with human subjects such as

---

* See Supplement F for a discussion of studies based on a single subject ($N = 1$) or very few subjects.

"placebo," "Hawthorne," or "Rosenthal" effects are particularly difficult to detect without control groups.

Sequential effects can be avoided by using matched pairs of subjects and testing each subject within a pair only once. Subjects must be matched on a variable relevant to the experiment. Certainly, matching subjects on the basis of height or weight would not be reasonable in a learning experiment, and matching on the basis of IQ scores would be of little value in a study of food consumption. The considerations involved in matching subjects are the same as those involved in deciding if a given sample is adequately representative. Matching subjects by using a pretest may well require as much work as using twice as many subjects in an independent-groups design. Matched groups, therefore, are used when the difficulty of matching is small compared to the difficulty in obtaining, training, or otherwise preparing the subject. Repeated testing on the same subjects is virtually unavoidable when the subjects are few and largely irreplaceable (e.g., captive killer whales).

Related-samples designs are also valuable when there are large differences among the subjects with regard to the characteristic being measured (the dependent variable). In the example of noisy and quiet working conditions, the effects of background noise may be relatively small compared to large differences in the ability of the subjects to solve the math problems. In related-samples tests we are concerned only with the differences in performance within each pair or the difference in the performance of each subject under the two conditions. Overall differences among subjects are essentially eliminated as a factor in the analysis.

A very efficient experimental procedure for dealing with situations involving large subject differences is the treatments-by-levels analysis-of-variance design, which will be discussed in Chapter 11. This design is based on the use of independent groups (but matched within levels) and statistically accounts for most of the differences among subjects so that the effect of the experimental variable is readily seen. The analysis also indicates if the effect of the experimental variable is different for the different types (levels) of subjects used.

WHITE TEST

Formula

T = sum of ranks of smaller group or, when $n_1 = n_2$, the smaller of the two rank sums.

$$T^1 = [n_1(n_1 + n_2 + 1)] - T$$

where n_1 is the size of the smaller group.

When $n_1 + n_2 > 30$, Table 1 cannot be used, and a normal distribution approximation is used where the T score for $p = .05$ and $p = .01$ is given by:

$$T = \left[\frac{n_1(n_1 + n_2 + 1)}{2}\right] - \left[Z\sqrt{\frac{n_1 n_2(n_1 + n_2 + 1)}{12}}\right]$$

where $Z = 1.96$ for $p = .05$
$Z = 2.58$ for $p = .01$

This approximation is often not suitable for $p = .001$.

Table I

provides probability values associated with T or T^1—whichever is smaller—when $n_1 + n_2 \leq 30$.

Table B

provides r_m values based on p and $(n_1 + n_2 - 2)$.

Procedure

1. Rank the scores across both groups.
2. Sum the ranks for each group to obtain T.
3. Calculate T^1.
4. Select the smaller of T and T^1.
5. Obtain p from Table I using n_1 to designate the size of the smaller sample and n_2 for the larger sample.
6. Obtain r_m from Table B using p and $(n_1 + n_2 - 2)$.

Example

The number of correct responses made by highly motivated subjects is compared to the number of correct responses made by poorly motivated subjects with the following results:

Highly Motivated Subjects	Rank	Poorly Motivated Subjects	Rank
20	9	10	5
16	8	7	3
15	7	5	1.5
12	6	5	1.5
9	4		
	34		$T = 11$

$T = 11$

$T^1 = [4(4 + 5 + 1)] - 11$

$\quad = 4(10) - 11 = 40 - 11 = 29$

T is smaller than T^1

From Table I ($T = 11$ when $n_1 = 4$, $n_2 = 5$),

$p = .05$

From Table B ($p < .05$ when $n_1 + n_2 - 2 = 7$),

$r_m = .67$

Conclusion

The highly motivated subjects make significantly ($p = .05$) and appreciably ($r_m = .67$) more correct responses than poorly motivated subjects.

MATCHED-PAIRS SIGNED-RANKS TEST

Formula

T = the sum of the ranks of the positive or negative differences, whichever is smaller.

When $N > 50$, use the normal curve approximation to obtain p from Table D based on Z where:

$$Z = \frac{T - \left[\dfrac{N(N-1)}{4}\right]}{\sqrt{\dfrac{N(N+1)(2N+1)}{24}}}$$

Table J

provides probability values based on T and N when $N \leq 50$.

Table B

provides r_m values based on p and $(N - 1)$.

Table D

provides probability values based on Z for the normal curve approximation.

Procedure

1. Obtain the difference between the two scores for each subject, keeping track of the direction of the difference (plus or minus).
2. Rank the differences from smallest to largest without reference to the sign of the difference.
3. Obtain the sum of the ranks of the plus differences and of the minus differences.
4. Select the smaller of the two sums.
5. Obtain p by entering T in Table J or use Table D when using the normal curve approximation for $N > 50$.

6. Obtain r_m based on p and $(N - 1)$ from Table B or, with the normal curve approximation, from Table C based on Z and N.

Example

Student performance on a mental agility test is measured at the start and the end of a course in basic statistics.

Subject	Before	After	Difference	Rank Plus	Rank Minus
a	80	95	+15	4	
b	70	60	−10		2
c	60	65	+5	1	
d	50	75	+25	7	
e	45	60	+15	4	
f	40	55	+15	4	
g	25	45	+20	6	
				26	2 = T

$T = 2$, since the sum of the ranks of the changes in the negative direction (after < before) is smaller than the sum for the positive direction.
From Table J (with $T = 2$, $N = 7$),

$p < .05$

From Table B (with $p < .05$, $N - 1 = 6$),

$r_m > .71$

Conclusion

The scores on the mental agility test were significantly ($p < .05$) and appreciably higher ($r_m > .71$) after the course in basic statistics than the scores before the course.

SUMMARY

1. *Categorical data* are obtained when observations are categorized, or separated, into discrete categories.

2. When the categories are numerous and relate to the degree or level of a characteristic, the data are in the form of *scores*.

3. Scores can be *ranked*, or ordered, according to their size, or value.

4. Scores that provide information about their relative position only and are treated in terms of their ranks, or order, are known as *ordinal data*.

5. *Random sampling*, which gives each member of the population an equal chance of being included in a sample, is used to produce unbiased samples.

6. The *median* is the score above and below which half the scores fall, that is, the middle score.

7. The *median test* is a simple but inefficient test for comparing the scores of two independent groups.

8. The *White test* compares the ranks of the scores of two independent groups and is more sensitive to the information provided by the scores than the median test.

9. *Related-samples designs* involve testing each subject under two or more conditions (repeated-measures design) or matching subjects in pairs or groups and testing one subject of a group under each condition (matched-group design).

10. The *sign test* is a generally weak test for data from two related samples based only on the direction of change from one condition to the other.

11. The *matched-pairs signed-ranks test* is a test for two related samples that uses both the direction and size of the changes and is more efficient than the sign test.

12. In studies using related samples, measuring each subject more than once in a *repeated-measures* design involves sequential, or carry-over, effects, which can affect the second and subsequent tests. Using *matched groups* permits each subject to be tested just once but involves problems with regard to obtaining properly matched groups. *Independent-samples* designs avoid these problems but generally require more subjects.

chapter five

KRUSKAL-WALLIS TEST FOR THREE OR MORE INDEPENDENT GROUPS

To study the effects of background noise level on students' problem-solving ability, we employ three conditions—quiet, slightly noisy, and very noisy—and obtain the results given in Table 5.1 in terms of the number of problems solved by each subject. The missing score in the group working under the slightly noisy condition is due to a subject who failed to appear for the experiment. In general, it is most efficient to have an equal number of subjects in each group. A large number of subjects in one group does not compensate for the unreliability arising from having too few in a second group.

TABLE 5.1. Number of Problems Solved by Each Student

Under Quiet Condition	Under Slightly Noisy Condition	Under Very Noisy Condition
18	14	8
17	12	6
15	9	5
15	9	4
10		2

If we treat the scores as ordinal data, by dealing with the ranks of the scores rather than the scores themselves, these data can be analyzed with an extension of the White test. If we rank the scores and then sum the ranks for each of the three conditions, the average rank for the three conditions should be virtually the same if the null hypothesis is true. This is the same rationale as that of the White test. We now need a statistic that gives a numerical value relating to the difference between the actual outcome and that expected if the three conditions were essentially the same. Although it is easy to obtain the difference between two sums of ranks, a slightly different procedure is needed to deal with differences among three or more sums.

Consider the following example of two numbers that always add up to 6, and see what happens when each number is squared and the squares are summed.

$$3^2 + 3^2 = 9 + 9 = 18$$
$$2^2 + 4^2 = 4 + 16 = 20$$
$$1^2 + 5^2 = 1 + 25 = 26$$
$$0 + 6^2 = 0 + 36 = 36$$

When a set of numbers adds up to the same final value, the sum of the squares of the numbers will increase as the initial values of the numbers become farther and farther apart. This fact provides a basis for measuring differences among three or more groups, and an appropriate statistic (H) has been developed by Kruskal and Wallis (1952):

$$H = \frac{12}{N(N+1)} \left(\sum \frac{R_j^2}{n_j} \right) - [3(N+1)] \tag{5.1}$$

where N = total number of subjects
  R_j = sum of ranks of a given group j
  n_j = number of subjects in group j

The sum of the ranks for each condition is squared and then divided by the number of scores in the column. This adjusts for any differences in the number of subjects in each condition. The value of the statistic will be the same whether we rank from small to high or from high to small, and therefore no adjustment for the direction of ranking, such as we use with the White test, is needed. When there are three or more conditions or 6 or more subjects in each group, the sampling distribution of H is essentially the same as the distribution of χ^2 with df = number of conditions $- 1 = k - 1$. The χ^2 approximation is somewhat inaccurate for three conditions with sample sizes no larger than 5, and exact values are given in Table K.

The scores for our example are ranked in Table 5.2. H is calculated as follows:

$$H = \frac{12}{14(15)} \left(\frac{58^2}{5} + \frac{32^2}{4} + \frac{15^2}{5} \right) - 3(15) = 10.65 \tag{5.2}$$

TABLE 5.2. Ranked Scores

Quiet Condition	Slightly Noisy Condition	Very Noisy Condition
14	10	5
13	9	4
11.5	6.5	3
11.5	6.5	2
8		1
$\Sigma = 58$	$\Sigma = 32$	$\Sigma = 15$

From Table K for $H = 10.65$, $p < .01$, and from Table B with $N = 14$ and $p < .01$, $r_m > .62$. There are large ($r_m > .62$) and significant ($p < .01$) differences in the scores among the three conditions, with the greatest number of correct responses having been obtained under the quiet condition and the fewest under the very noisy condition.

As with χ^2 for more than two conditions, the interpretation of the results is not always simple. The significant and appreciable H value in the example means basically that we cannot assume the three sets of subjects are drawn from the same population; that is, the three experimental conditions are not equivalent. Can we say anything more about the data? When the experimental conditions fall on a continuum, as in this example, the interpretation is fairly easy. It is, generally, sufficient to order the conditions in terms of the dependent variable, that is, to indicate which condition gave the highest scores and which condition gave the lowest. This does not mean, however, that we can say there are significant differences between any two of the conditions, since this was not specifically tested. The **Kruskal-Wallis test** is concerned with whether all the samples can be assumed to be random samples drawn from the same population and unaffected by the independent variable.

If the Kruskal-Wallis test indicates further comparisons are desirable, we can compare two of the conditions by using the appropriate test for two conditions—the White test. In this example there is a moderate, but not significant, difference between the quiet and slightly noisy conditions ($p > .05$, $r_m < .67$). Both the quiet and slightly noisy conditions are appreciably and significantly different from the very noisy condition ($p < .05$, $r_m > .63$). This indicates that a slightly noisy condition causes, at most, a slight deterioration in performance compared to the quiet condition, but, relative to either the quiet or slightly noisy condition, a very noisy condition causes a marked drop in performance.

This procedure of making specific comparisons between pairs of conditions is subject to certain statistical problems. Basically, the more comparisons made, the greater is the likelihood of obtaining a significant difference by chance—in other words, of committing a type I error. For example, if there are 7 groups giving $(7 \times 6)/2 = 21$ possible comparisons of all pairs, by chance alone we would expect at least 1 of these comparisons (1/20) to be significant at the .05 level. There are statistical procedures that adjust the probability levels so as to account for the number of possible comparisons (cf. Ryan 1962; Kirk 1968). It is usually safe to compare only those conditions that are reasonable in terms of the independent variable, and not to compare groups primarily because their scores appear to be different. Furthermore, it is best to be a bit conservative (use a more stringent level of significance such as .01 instead of .05) in rejecting the null hypothesis when making these multiple comparisons.

The Kruskal-Wallis test is convenient for screening a large number of conditions to see if there might be any differences among the conditions. If there is no significant or appreciable overall effect, there is no need to compare specific pairs of conditions. Here too we would have to take sample size and magnitude of effect into account in deciding whether or not observed differences are worth considering further.

A multiple-group study may contain a natural reference group. An example might be the testing of several different drugs with a control group getting only a saline injection. In that case we would be primarily interested in comparing each drug group against the control group and less interested in comparing one drug group to another. In this type of situation, which is fairly common, we would be reasonably safe in using a two-sample test to compare each drug to the control.

TABLE 5.3

A	B	C
1	2	3
6	5	4
7	7	7

If the sums of ranks for all the columns are equal and there is an equal number of subjects in each column, as shown in Table 5.3, if we use formula (5.1), the value of H will equal 0:

$$H = \frac{12}{6(7)} \left(\frac{7^2}{2} + \frac{7^2}{2} + \frac{7^2}{2} \right) - 3(7)$$

$$\frac{2}{7}(73.5) - 21 = \frac{147}{7} - 21 = 21 - 21 = 0 \tag{5.3}$$

In other words, when the outcome is exactly as would be expected by chance, the value of the statistic is 0, and the statistic becomes larger as the differences among the column ranks become larger. This statistic, then, has the same general characteristics as χ^2.

The Kruskal-Wallis test is efficient and readily calculated when the samples are small. As indicated by Table K, this test is sufficiently sensitive to enable us to reject the null hypothesis with sample sizes as small as 3/3/2. However, ranking scores across several groups, when the samples are large, is an awkward procedure and subject to errors.

FRIEDMAN TEST FOR THREE OR MORE RELATED SAMPLES

Suppose we wish to investigate the effects of background noise on students' problem-solving ability, as in the previous section, but in this instance it appears that the effects of the noise may be obscured by large differences in the natural ability of the subjects. We decide to use a repeated-measures design, testing each subject under each of the three noise conditions. An alternative method would be to use matched groups of subjects, with one member of each group being tested under each condition. Since we are also concerned about possible sequential effects such as practice or boredom, we give several short test sessions under each of the three conditions—quiet, slightly noisy, and very noisy. The measure is the total score for each subject under each of the three conditions. The order in which the students are subjected to the three conditions is counterbalanced so that any sequential effects will contribute equally to all three conditions. With the total number of problems solved as the measure, we obtain the results given in Table 5.4.

TABLE 5.4

Subject	Quiet Condition	Slightly Noisy Condition	Very Noisy Condition
a	22	16	14
b	15	12	8
c	10	6	7
d	8	5	3

Since each subject is tested under each of the three conditions, there is an equal number of scores under each condition. We now need a statistic that will measure the degree to which these three sets of scores differ. To handle data from more than two samples, the Kruskal-Wallis test used an extension of the two-sample, or White, test. Thus it would seem reasonable to use an extension of the matched-pairs signed-ranks test to handle more than three conditions with related samples. The matched-pairs test, however, depends on getting differences between two scores for each subject, and this is not feasible for more than two test conditions. The appropriate procedure is an extension of the sign test. Although the sign test is weak and inefficient when used with two groups, it is surprisingly effective for three or more groups. To test the performance of the individual subjects under the three conditions, each subject's three scores are ranked, from either lowest to highest or highest to lowest, across the three conditions. These ranks are summed across subjects for each of the three conditions, as in the Kruskal-Wallis test, and the sum of the square of the ranks provides a measure of the differences among the column sums. An appropriate statistic for this test (χ_r^2) was devised by M. Friedman* (1937) and has the following form:

$$\chi_r^2 = \left[\frac{12 \sum (R_j^2)}{Nk(k + 1)}\right] - 3N(k + 1) \tag{5.4}$$

where N = total number of subjects
        k = number of groups
        R_j^2 = the square of the sum of ranks for group j

TABLE 5.5

Subject	Quiet Condition	Slightly Noisy Condition	Very Noisy Condition
a	1	2	3
b	1	2	3
c	1	3	2
d	1	2	3
	4	9	11

For example, the scores given in Table 5.4 are ranked in Table 5.5, and χ_r^2 is calculated as follows:

$$\chi_r^2 = \frac{12(4^2 + 9^2 + 11^2)}{4(3)(4)} - 3(4)(4) = \frac{(218)}{4} - 48$$

$$= 54.5 - 48 = 6.5$$

$$p < .05 \qquad r_m > .53 \tag{5.5}$$

There are significant and fairly large differences in performance under the three noise conditions with the best performance being obtained under the quiet condition and the worst performance under the noisy condition. As with the statistic for the Kruskal-Wallis test, the formula for χ_r^2 for the Friedman test gives a value of 0 when the ranks of the three conditions all have the same sum. As the difference between scores in the three columns increases, the value of the statistic gets larger. With large sample sizes the probability values associated with this statistic are approximated by the χ^2

* No known relation but, doubtless, a capable and charming fellow.

distribution with $df =$ conditions $- 1 = k - 1$. For small sample sizes ($N < 10$, $k = 3$; or $N < 5$, $k = 4$), the χ^2 approximation is slightly inaccurate, and the exact probability values are given in Table L.

With only two conditions the assignment of ranks 1 and 2 would be equivalent to the $(+)$ or $(-)$ categories used in the sign test, and the two tests would be identical. With three or more conditions this procedure is no longer inefficient, and the Friedman test turns out to be both sensitive and easy to use.

The interpretation of the data is essentially equivalent to that used for the Kruskal-Wallis test. When it is desirable to compare two of the experimental conditions, the appropriate two-sample test—the Wilcoxon signed-ranks test —is used. In the present example, the sample size is too small to use the Wilcoxon test. The same problems and precautions hold here as for the Kruskal-Wallis test when making multiple comparisons of pairs of conditions after the overall Friedman test.

KRUSKAL-WALLIS TEST

Formula

$$H = \frac{12}{N(N + 1)} \left(\sum \frac{R_j^2}{n_j} \right) - [3(N + 1)]$$

where n_j = size of sample j

$N = \sum (n_1 + n_2 + n_3 \cdots)$ = number of subjects in all samples combined

R_j = sum of ranks of scores in sample j

Table K

provides p values based on H and n_j for three groups with groups no larger than $n_j = 5$.

Table F

provides p values based on H (treated as χ^2) and df = number of groups $-$ 1 when $n_j > 5$ or four or more samples are involved.

Table B

provides r_m values based on p and N.

Procedure

1. Rank the scores (preferably from low to high) across all groups.
2. Sum the ranks for each group to obtain R_j and square the term to obtain R_j^2.
3. Divide the R_j^2 term for each group by the number of subjects in that group (n_j) to obtain R_j^2/n_j.
4. Calculate $\sum (R_j^2/n_j)$ by adding the terms from step 3.
5. Determine the value of H by using the formula above.
6. Use Table K to obtain p values based on H and the size of each sample or Table F for p values based on H (as χ^2) and df = number of samples $-$ 1.
7. Use Table B for r_m values based on p and N.

Example

Three groups of rats are tested in a Skinner box, with the rate at which the bar is pressed for food being the measure. Each group is run under a different level of food deprivation.

Number of Responses in Five Minutes

Under Slight Deprivation, $n_1 = 4$	Rank	Under Moderate Deprivation, $n_2 = 4$	Rank	Under Extreme Deprivation, $n_3 = 3$	Rank
2	1	7	4	25	8
4	2.5	15	6	40	10
4	2.5	20	7	50	11
8	5	30	9		
$\sum R_1 = 11$		$\sum R_2 = 26$		$\sum R_3 = 29$	

$$N = n_1 + n_2 + n_3 = 4 + 4 + 3 = 11$$

$$H = \frac{12}{11(12)}\left[\frac{(11)^2}{4} + \frac{(26)^2}{4} + \frac{(29)^2}{3}\right] - 3(12)$$

$$H = \frac{1}{11}\left[\frac{121}{4} + \frac{676}{4} + \frac{841}{3}\right] - 36$$

$$H = \frac{479.58333}{11} - 36 = 43.5985 - 36 = 7.5985$$

From Table K, $p < .01$ based on $H = 7.5985$ for samples of 4, 4, and 3.
From Table B, $r_m > .68$ based on $p < .01$ and $N = 11$.

Conclusion   There are significant ($p < .01$) and moderately large differences ($r_m > .68$) in the response rates among the three groups of rats. The extremely deprived group had the highest rate of response and the slightly deprived group had the lowest.

FRIEDMAN TEST

Formula

$$\chi_r^2 = \frac{12 \sum (R_j^2)}{Nk(k+1)} - 3N(k+1)$$

where N = number of subjects or matched groups
k = number of conditions
R_j^2 = the square of the sum of ranks for group j

Table L

provides p values based on χ_r^2, k, and N when there are 3 conditions with no more than 9 subjects or 4 conditions with no more than 4 subjects.

Table F

is used when k or N is too large for Table L and provides p values based on χ_r^2 treated as χ^2 with $df = k - 1$.

Table B

provides r_m values based on p and N (k).

Procedure

1. Rank the scores of each subject or matched group across the k conditions (i.e., from 1 to k) preferably from lowest to highest.
2. Sum the ranks of the scores in condition j. Square this sum to obtain (R_j^2).
3. Repeat step 2 for all conditions and add the (R_j^2) terms to obtain $\sum (R_j^2)$.
4. Calculate the value of χ_r^2.
5. Enter χ_r^2, N, and k in Table L to obtain p values.
6. If $k = 3$ and $N > 9$, $k = 4$ and $N > 4$, or $k > 4$, enter χ_r^2 as χ^2 in Table F with $df = k - 1$ to obtain the p value.
7. Enter p and N (k) in Table B to obtain the r_m value.

Example   To test for the effects of practice on students' ability to solve simple problems, three two-minute math tests were administered to each subject. The problems on each test were of equivalent difficulty, and the second and third tests were preceded by five-minute rest periods. The measure was the number of problems solved on each test.

Subject, $N = 5$	Test 1	Rank	Test 2	Rank	Test 3	Rank
a	5	1	7	2	10	3
b	4	1	9	2	11	3
c	8	2	5	1	12	3
d	6	1	15	3	11	2
e	9	1	12	2	16	3
		$\sum R_1 = 6$		$\sum R_2 = 10$		$\sum R_3 = 14$

$$\chi_r^2 = \frac{12}{5(3)(3+1)} (6)^2 + (10)^2 + (14)^2$$
$$- 3(5)(3+1)$$
$$= \frac{332}{5} - 60 = 66.4 - 60 = 6.4$$

From Table L, $p < .05$ for $\chi_r^2 = 6.4$, $N = 5$, and $k = 3$. From Table B, $r_m > .48$ for $p < .05$ and $N(k) = 15$.

Conclusion   The number of problems solved differs significantly ($p < .05$) and moderately ($r_m > .48$) across the three test sessions with the most problems solved in the third session and fewest in the first.

SUMMARY

1. The *Kruskal-Wallis test* for three or more independent groups, or samples, is an extension of the White test and, therefore, deals with the ranks of the scores.

2. The *Friedman test* for three or more related samples is a sensitive and efficient extension of the sign test.

chapter six

RANK-ORDER CORRELATION

The tests discussed so far have been directed at analyzing differences, either among groups or between the data and some prior expectation. In many situations questions about differences are not the only ones that may be asked. It is often interesting and valuable to determine how well certain variables are related to each other. For example, are grades related to IQ? This chapter is concerned with measuring relationships between variables.

In order to find the relationship between height and weight in men, we select a representative sample of the population and measure the height and weight of each subject. The data are shown in Table 6.1. We now need a statistic that describes the degree of **correlation** between height and weight— that is, a statistic that indicates how these two variables are related to each other, or correlated.

TABLE 6.1

Subject	Height (inches)	Weight (pounds)
a	60	100
b	63	155
c	65	170
d	68	150
e	70	175
f	73	265
g	78	237

For the moment we are concerned with only the ordinal, or ranking, properties of the measurement. In a later chapter comparable procedures that deal with the numerical value of the scores will be described. We can consider the two variables (height and weight) as being related if an individual's score or rank on one variable (weight) can be predicted, or estimated, on the basis of his score or rank on the other variable (height). The question can be stated as: Does the individual's relative standing (i.e., his rank) within the sample in terms of weight correspond to his relative standing, or rank, in terms of height? If the two sets of rankings are identical, with the tallest person being the heaviest, the shortest being the lightest, and correspondingly in between, then the correlation is a perfect one, and we can accurately indicate an individual's rank with regard to weight if we know his ranking with regard to height. At the other extreme, if the two sets of rankings bear no correspondence and the taller individuals have both low and high rankings in terms of weight, then knowing an individual's height is of no help in estimating his weight. The **Spearman rank-order correlation** (r_s) is the statistic that gives a numerical value to the degree of correspondence between the two sets of rankings. It is derived from the correlation measure used for scores (Chapter 8) and has the following form:

$$r_s = 1 - \left[\frac{6 \sum (d^2)}{N(N^2 - 1)} \right] \tag{6.1}$$

where d = the difference in ranks for each subject
N = number of subjects

Basically the operation compares each individual's rankings on the two variables and gives the difference between the two sets of rankings. Clearly, if the rankings of both variables are identical, the term $\sum (d^2) = 0$, and $r_s = 1 - 0 = 1.0$.

TABLE 6.2

Subject	Height	Rank H	Weight	Rank W	Difference (d) in Ranks ($H - W$)	d^2
a	60	1	100	1	0	0
b	63	2	155	2	0	0
c	65	3	170	4	−1	1
d	68	4	160	3	+1	1
e	70	5	175	5	0	0
f	73	6	265	7	−1	1
g	78	7	237	6	+1	1
						$\sum (d^2) = 4$

For the example of Table 6.1 the data are ranked in Table 6.2, and the statistic is calculated like this:

$$r_s = 1 - \frac{6(4)}{7(49 - 1)} = 1 - \frac{24}{336} = 1 - .07 = .93 \qquad (6.2)$$

Next we have to interpret the correlation coefficient just calculated, but as a first step in understanding the value of this statistic, we will consider a simple but extreme situation, which illustrates certain other aspects of the correlation coefficient, before returning to the height-weight data.

SCATTERGRAM

In a trivial study of the relationship between the number of problems answered correctly (C) and the number answered incorrectly (IC) on a 20-question quiz, we obtain the data in Table 6.3 and the following correlation coefficient:

$$r_s = 1 - \frac{6(40)}{5(25 - 1)} = 1 - \frac{240}{120} = 1 - 2 = -1 \qquad (6.3)$$

In terms of our ability to predict an individual's rank on one variable from his rank on another variable, it is clear that there is a perfect relationship—though a negative one. The degree of relationship is the same when $r_s = 1$ and when $r_s = -1$. The sign of the correlation indicates whether high rankings on one variable are associated with high rankings on the other (a positive correlation) or high rankings on one variable are associated with low rankings on the other variable (a negative correlation).

TABLE 6.3

Subject	Number Correct (C)	Rank C	Number Incorrect (IC)	Rank IC	d	d^2
a	20	5	0	1	4	16
b	16	4	4	2	2	4
c	12	3	8	3	0	0
d	9	2	11	4	2	4
e	6	1	14	5	4	16
						$\sum d^2 = 40$

This is easier to see if the data given in Table 6.3 are graphed, with the two rankings for each individual locating the points on the graph, as in

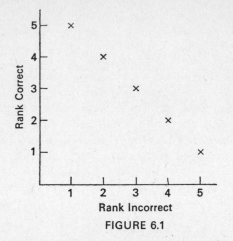

FIGURE 6.1

Figure 6.1. A graph that shows the scatter of the data points is known as a scatter diagram, or **scattergram**. In this example a straight line can be drawn through all of the points. As will be discussed further in Chapter 8, the degree of relationship is indicated by the degree to which the points on the scattergram approximate a straight line. The scattergram in Figure 6.1 shows a perfect correlation between the two variables. The slope of the line, from the upper left to the lower right, shows a negative relationship in that high values of one variable tend to be associated with the low values of the other variable.

The scattergram for the data of heights and weights given in Table 6.2 looks like that shown in Figure 6.2. In this case the points do not all fit on a straight line, reflecting that the correlation is less than a perfect one. However, they fall about a straight line (which is drawn in lightly), and this

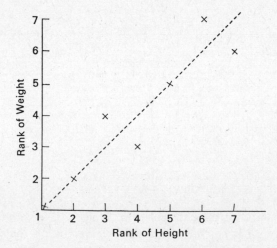

line goes from the lower left to the upper right with the high ranks on one variable tending to be associated with high ranks on the other variable, showing a high positive relationship.

If we consider yet another example, the relationship between weight and IQ, we might get the type of data illustrated in Figure 6.3. Here there is essentially no relationship between the two variables, and a correlation coefficient based on these data would be virtually 0.

The value of the correlation coefficient, therefore, ranges from 0 for no relationship to a maximum of 1.0 for a perfect relationship. The sign of the correlation (+ or −) indicates the *direction* of the correlation and has nothing at all to do with the *degree* of correlation. Therefore, a correlation of −.6 represents a greater relationship than does a correlation of +.4.

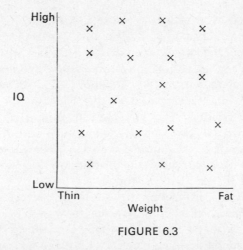

FIGURE 6.3

LINEAR OR CURVILINEAR RELATIONSHIPS

The scattergram can provide another bit of information in addition to the direction of the relationship and the approximate value of the correlation coefficient. Keeping in mind that the correlation coefficient is a measure of the degree to which the points of the scattergram fit on a straight line, consider the following situation. We are interested in the relationship between the number of hours rats are deprived of food and the rate at which they press a bar to obtain food. The data yield the scattergram of Figure 6.4. Notice that initially the rate of response increases with increasing deprivation, but as deprivation continues, the rate of response declines and finally drops to 0. At this point (120 hours), the astute experimenter would have noticed the subject's general lack of activity (except perhaps, aromatic). Clearly no straight line would fit these points; yet the relationship between deprivation and rate of response is apparently a strong one. The value of r_s calculated

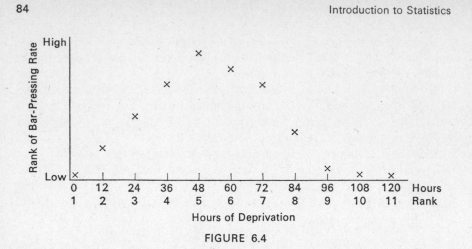

FIGURE 6.4

from these data would be virtually 0. The relationship is not a **linear** one, which could be approximated by a straight line, but rather a **curvilinear**, or curved line, relationship, which can only be described with a nonlinear function, or curve. There are correlation techniques beyond the scope of this text that can be used to analyze this situation. The scattergram, then, indicates whether or not the linear correlation measure is appropriate. Meaningful rank correlations can be obtained from slightly nonlinear distributions, but a markedly curvilinear function, such as in the example of Figure 6.4, is inappropriate for a linear correlation. The scattergram of the height-weight data (Figure 6.2) shows the relationship to be linear as well as high and positive.

We now need to consider whether or not a given correlation coefficient might represent a chance relationship between the variables. Consider the following situation: In a test for the possible correlation between weight and IQ, the first randomly selected subject is both rather portly and unusually intelligent. The second subject turns out to be slight in both build and mental endowment. A rapidly constructed scattergram gives the picture shown in Figure 6.5. With both points on the straight line, the correlation is clearly a perfect, positive one ($+1.0$), but, then again, any two points can be connected by a straight line. (This is true even if you do not remember your plane geometry.) The calculations for r_s [formula (6.1)] show that, no matter what the scores are, when there are 2 individuals, the rankings will give either a $+1.0$ or -1.0 value for r_s. Therefore, a perfect correlation coefficient obtained with 2 subjects ($N = 2$) is trivial. What happens when we measure a third subject? His score does not have to fall on the straight line; in fact we would be surprised if it did. As more and more subjects are measured, it becomes increasingly unlikely that the points will fall exactly on or very near to the initial straight line unless there is a relationship between the 2 variables.

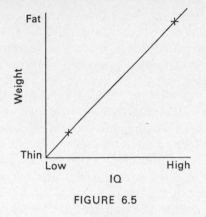

FIGURE 6.5

We can construct a sampling distribution for r_s, though a different one is needed for each sample size. For example, a sampling distribution for $N = 10$ would be obtained by randomly selecting 10 pairs of scores from a table of random numbers and obtaining the correlation coefficient. Repeating the process, we would eventually end up with a distribution of all the chance values of r_s. The distribution would have the most scores at or about 0 and a decreasing number of scores extending toward $+1$ and -1. Values from mathematically calculated sampling distributions for the r_s statistic for the .05 and .01 probability levels are given in Table M. There is no point in calculating the r_m value for the correlation, since the degree of relationship between the two variables is given directly by r_s. In the example for height and weight (Table 6.2 and Figure 6.2), the value of r_s ($+.93$) for $N = 7$ is significant with $p < .01$. Therefore, the relationship between height and weight is linear, yielding a high, positive ($+.93$), significant ($p < .01$) correlation, with large height tending to be associated with high weight.

INDEPENDENT AND DEPENDENT VARIABLES— EXPERIMENTS VERSUS CORRELATIONAL STUDIES

We can now broadly distinguish between the two basic types of scientific investigations. In an **experiment**, one variable, the **experimental**, or **independent**, variable is varied, or manipulated, and some other aspect of the situation, the **dependent** variable, is measured. Often it is impractical to manipulate variables. (We do not change the weight of subjects to see the effect upon height, nor can we force people to smoke cigarettes and then wait to see how long they live.) In this case the appropriate procedure is to use a **correlational study** and measure both variables of interest. When we are dealing with people, particularly with personality or health factors, experiments are often inappropriate, but correlations can be very informative. Quite often variables are first detected as being important in correlation studies and are further investigated by experimental manipulation. In Chapter 12 we will see that the r_m value is basically a correlational measure. Therefore, the interpretation of both experiments and studies is based on the use of probability and correlational measures.

The most important difference between experimental and correlational studies is that causal relationships between variables can be seen clearly in an experiment only. Manipulating one variable and observing the effect on another variable provides an unambiguous test of whether or not the first variable can affect, or cause changes in, the other. Correlations between variables show only that the two variables are associated, not that one causes changes in the other. The problem of attributing causality on the basis of correlational data will be discussed in Chapter 8.

RANK-ORDER CORRELATION

Formula
$$r_s = 1 - \left[\frac{6 \sum (d^2)}{N(N^2 - 1)}\right]$$

Table M   provides p values based on r_s and N

Restrictions   The rank-order correlation is a measure of *linear* relationship. If the relationship is essentially nonlinear or curvilinear, as shown by a scattergram, this measure will not provide a meaningful view of the data.

Procedure
1. Rank the scores separately for each measure (preferably from low to high) across all subjects.
2. For each subject obtain the difference (d) between the ranks of the two measures for that subject.
3. Square the difference for each subject to obtain (d^2).
4. Add up the (d^2) values to obtain $\sum (d^2)$.
5. With N = number of subjects, calculate ($N^2 - 1$) and multiply by N to obtain $N(N^2 - 1)$.
6. Divide $6 \sum (d^2)$ by $N(N^2 - 1)$ and obtain r_s.
7. The *direction* of the correlation is given by the sign of r_s (+ or −), and the *degree* of the relationship is given by the value of r_s.
8. Obtain the p value associated with r_s from Table M.

Example   The College Board verbal and quantitative scores for a group of students are:

Subject	Verbal Score	Rank	Quan- titative Score	Rank	d	d^2
a	320	1	410	2	−1	1
b	400	2	380	1	+1	1
c	470	3	460	3	0	
d	510	4	530	5	−1	1
e	540	5	490	4	+1	1
f	590	6	660	7	−1	1
g	630	7	670	8	−1	1
h	680	8	620	6	+2	4
						10

$$r_s = 1 - \left[\frac{6(10)}{8(64 - 1)} \right] = 1 - \left[\frac{60}{504} \right] = 1 - .1190$$

$$= +.88$$

From Table M ($r_s = .88$ with $N = 8$),
$p < .01$

Conclusion   There is a significant ($p < .01$) and very high positive correlation ($r_s = +.88$) between ranked verbal and quantitative College Board test scores. High verbal score ranks tend to be associated with high quantitative score ranks.

SUMMARY

1. The *rank-order correlation* (r_s) is a measure of the degree to which the ranking of individuals according to one variable agrees with their ranking on another variable. When the correlation is perfect, $r_s = 1.0$; when there is no relationship, $r_s = .00$. The *sign* of the correlation is positive ($+$) when high ranks on one variable tend to correspond with high ranks on the other and negative ($-$) when high ranks tend to correspond with low ranks. The sign is independent of the degree of the correlation coefficient.

2. A *scattergram*, or scatter diagram, gives a visual representation of correlation data in graph form. Each subject is represented at the point on the graph where his two scores (one on each coordinate) intersect. The scattergram permits an estimation of the degree, or size, and direction, or sign, of the relationship.

3. The rank-order correlation measure is suitable only for relationships that can be approximated by a straight line on the scattergram—that is, *linear relationships*. A *curvilinear relationship*, one that can be represented by a curved line, can only be measured with more complex correlation procedures.

4. The *independent*, or *experimental*, *variable* is the variable or aspect of an experiment that is varied, or changed, for the different groups of subjects, while another variable, the *dependent variable*, is measured. Both variables in a correlational study are, in effect, dependent variables.

5. In an *experiment* one or more aspects of a situation are varied, and the effect on the dependent variable is measured. Since the manipulations of the independent, or experimental, variable precede the measurement of the dependent variable, any causal relationship between the independent and dependent variables is apparent. In a *correlational study*, neither variable is manipulated, and causal relationships *cannot* be assumed on the basis of an obtained correlation.

chapter seven

CATEGORICAL, ORDINAL, AND INTERVAL DATA

The previous chapters dealt with categorical data and ordinal data. A third type of data, **interval** data, deals directly with the numerical values of the scores. In other words the intervals between the scores, not just the order of the scores, are taken into account. The differences among these three types of data (cf. Stevens 1946) can now be summarized.

Assume that we measure the heights of a group of individuals and divide the subjects into two categories: tall and short (Figure 7.1). In doing this, we are using only the categorical information. If we go further and order the subjects on the basis of height from shortest to tallest, we are making use of the *ordinal* properties of the data (Figure 7.2). Finally, if, in addition to ordering the subjects, we use the size of the differences, or intervals, between any two scores in order to *space* the subjects according to their height, then we are making use of the *interval* properties of the data (Figure 7.3).*

In order to extract the maximum amount of information, it is often best to treat data as interval data—but only if this would not be misleading. Most

---

* Stevens discussed an additional type of measurement, which has all of the properties of an interval scale with the addition of an absolute zero point. With this scale, a *ratio scale*, not only do equal-sized intervals, or differences between scores, represent equal differences, but the ratio of two scores is a meaningful measure. For example, in a centigrade temperature scale (an interval scale with an arbitrary zero point) the difference between 10° C. and 20° C. represents the same energy difference as between 40° C. and 50° C., but water is not *twice* as hot at 20° C. as at 10° C. With a Kelvin scale having an absolute zero point (a ratio scale), 300° K. is twice as hot as 150° K. Physical measurements such as height and weight are common examples of ratio scales. This text is not concerned with tests specifically for use with ratio data and uses interval data tests for these situations.

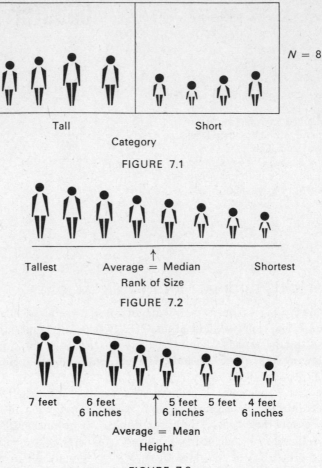

FIGURE 7.1

FIGURE 7.2

FIGURE 7.3

tests for interval data involve certain assumptions concerning the populations from which the samples were drawn, and frequently these assumptions are not supported by the data, in which case it may be best to treat the data in terms of their ordinal properties. In practice there is generally very little difficulty in deciding whether to treat scores in terms of their ordinal or interval properties (cf. Chapter 9).

CENTRAL TENDENCY

Mean

To deal with interval data, we first need a measure of **central tendency**—that is, a single score to represent the entire sample. If you had three exam grades (e.g., 4, 8, and 9), what would your average score be? The obvious solution is to add up the three numbers $(4 + 8 + 9 = 21)$ and divide this sum by the number of test grades $(21/3 = 7)$. The average score for the three grades is, therefore, 7. This average is called the **mean** and is defined by the following formula:

$$\text{mean} = \bar{X} = \frac{\sum X}{N} \tag{7.1}$$

where X = each score
N = the total number of scores

This formula is an exact description of the procedure used for computing the mean of the three test grades; the scores were added and divided by the number of scores.

Median

The average appropriate for ordinal data is the score of the subject in the middle when the scores are ranked—that is, the score associated with the middle rank. In other words, it is the score that divides the sample into two halves, and as we saw in the median test, this score is the *median*.*

The mean has several properties that are worth discussing. Look at Table 7.1, which gives the deviation of each score from the mean $(X - \bar{X})$ for the

TABLE 7.1

X	$\bar{X}$	$X - \bar{X}$
4	7	-3
8	7	$+1$
9	7	$+2$
		$\sum = 0$

test-score example. As in the table, the sum of the deviations about the mean is always equal to 0. In that sense the mean is centrally placed among the distribution scores. As an analogy, consider the balancing of a ruler with equal weights placed at the 4-, 8-, and 9-inch positions; the point of balance

---

* Note that the words "mean" and "median" both refer to being in the middle or being of medium or middling quality. The word "mediocre" comes from the same root and has the same meaning. In recent years, terms like "average" and "mediocre" have somehow acquired a pejorative connotation so that to be "merely average" is now (illogically) synonymous with "substandard." A husband who describes his wife's latest cake as "As good as you always make it, Honey!" is wise. One who describes it, equivalently, as "Mediocre, just average for you, Dear" is a fool.

is at the 7-inch mark. The plus weights, or weights pushing down on one side, counterbalance exactly the minus weights, those pushing down on the other side, and the ruler balances—that is, the plus and minus weights (or deviations) cancel out and sum to 0.

Method of Least Squares

We can go a step further and square the deviations about the mean and add these up. We can also square the deviations about some number other than the mean and see what the sum is (Table 7.2). The sum of the squared

TABLE 7.2

X	$X - \bar{X}$	$(X - \bar{X})^2$	X	$X - 6$	$(X - 6)^2$
4	-3	9	4	-2	4
8	$+1$	1	8	$+2$	4
9	$+2$	4	9	$+3$	9
$\bar{X} = 7$		$\sum = 14$	$\bar{X} = 7$		$\sum = 17$

deviations about the mean will always be smaller than the sum of the squared deviations about *any* other point in the distribution. The mean may be viewed as determined by the **method of least squares**, and the mean is the best fitting or the most representative point of the distribution in the sense that the sum of the squares of the deviations about this point is less than about any other point. The use of the method of least squares is going to appear again with regard to measuring relationships between two sets of measures (Chapter 8).

Mode

The mean is generally the most suitable measure of central tendency and is less disturbed by chance sampling factors than is the median. A third measure of central tendency is the score that occurs most frequently in the sample, or the **mode**. It is highly subject to chance factors and difficult to interpret and therefore is rarely used, except for categorical data when the categories are discrete and cannot be ordered.

SKEWED AND SYMMETRICAL DISTRIBUTIONS

Under certain circumstances, the mean can be a misleading measure of central tendency. Consider the distribution of incomes in a small mill town, as shown in Figure 7.4. If you were thinking of setting up an automobile agency in this town, the mean income ($18,000) would lead you to expect

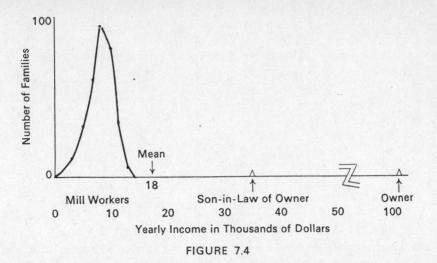

FIGURE 7.4

that the less ostentatious luxury cars might enjoy good sales. A closer look at the distribution reveals that there is *no one* in the town earning anywhere near $18,000. The mean, the most representative score, appears to be representative of none of the sample. What has gone wrong? The distribution obviously has two extremely high scores, which exert a disproportionately large influence upon the mean (in line with the balancing property of the mean). A distribution having this shape is known as a **skewed** distribution. A skewed distribution, one that is not symmetrical about the mean, can be misinterpreted if the mean is used as a measure of central tendency. Even if the distribution is skewed, the mean can still be determined by the method of least squares, and the sum of the deviations about the mean is equal to 0. With the incomes in this mill town, the automobile dealer would be better off selling low-priced cars, which are more appropriate to the median income. In a **symmetrical** distribution, a nonskewed distribution that has the same shape on both sides of the mean, the mean and median, will coincide. Therefore, the degree of disparity, or difference, between the mean and the median for a given distribution can be taken as a rough measure of the degree to which the distribution is skewed. In highly skewed distributions, the median is often a more reasonable and appropriate measure of central tendency than the mean.

STANDARD DEVIATION

The distributions of intelligence scores, as measured by an IQ test, are given for two colleges in Figure 7.5. The mean IQ for each college is 120. On the basis of the identical mean values, can we say that the two distributions are equivalent? Obviously not, since the spread of scores in College B is much greater than that in College A. Such differences might arise in the

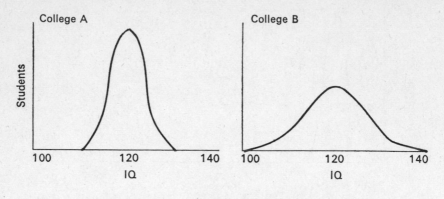

FIGURE 7.5

following way: College A is a second-rate institution with unrealistic and pretentious standards that lead it to reject applicants of marginal ability, but, in turn, it is avoided by students of appreciable ability. The net result is a rather truncated spread or limited variability of IQ scores. College B, on the other hand, is a large urban school with a good reputation, which attracts very good students while also admitting promising students with limited academic background. The difference between these two distributions is not in terms of the measure of central tendency (the mean) but rather in terms of the spread of the scores about the mean. Now we have to develop a statistic that will provide a satisfactory measure of variability.

For the test scores given earlier in the chapter, the most obvious measure might appear to be the sum of the deviations about the mean. But, as shown in Table 7.1, this sum will always be 0, regardless of the spread of the scores about the mean, since the $(+)$ and $(-)$ differences around the mean cancel out. We could arbitrarily use the absolute values of the differences, disregarding sign, and add these up, but this is not algebraically permissible and would limit the utility of the statistic.

There is another way of coping with this problem, one that was used previously in developing the χ^2 statistic. The squares of the deviations will all be positive regardless of the sign of the deviation, and therefore we can add the squares without their canceling (Table 7.3). So far, so good. But

TABLE 7.3

X	$\bar{X}$	$(X - \bar{X})$	$(X - \bar{X})^2$
4	7	-3	9
8	7	$+1$	1
9	7	$+2$	4
			$\Sigma = 14$

what would happen if there were six exam grades instead of three? With six scores being added, the value of $\sum (X - \bar{X})^2$ would be larger than with three even if the spread of scores were the same. Yet the statistic should not be directly influenced by the size of the sample, or the number of subjects, used. The solution is simple: We divide $\sum (X - \bar{X})^2$ by the number of scores (N) in order to obtain an average, or mean value, of the squared deviations. This statistic would serve fairly well, but for convenience and reasons that will become apparent later in this chapter, the square root of this term is usually used. This statistic is the most commonly used **measure of dispersion**, or variability. In fact, it is the standard measure of deviation about the mean and is called the **standard deviation**. The standard deviation (SD) is usually symbolized by a lower case sigma (σ) and has the following formula:

$$\text{SD} = \sigma = \sqrt{\frac{\sum (X - \bar{X})^2}{N}} \qquad (7.2)$$

where X = a given score
$\bar{X}$ = the mean

The test scores in the example lead to the following calculation:

$$\sigma = \sqrt{\frac{14}{3}} = \sqrt{4.67} = 2.16 \qquad (7.3)$$

The standard deviation does exactly what we want. It reflects the spread of the scores about the mean, taking into account the number of scores in the sample, and it is a fairly stable measure that is minimally affected by chance fluctuations in the composition of the sample. Whenever the mean is an appropriate measure of central tendency, the standard deviation is the appropriate measure of dispersion.

BIASED AND UNBIASED ESTIMATES

The standard deviation does involve one minor problem, but we can deal with this quite simply. Whenever the sample size is less than infinitely large (a surprisingly common occurrence), the standard deviation of the sample tends to be *biased* in the direction of *underestimating* the SD of the population from which the sample was drawn. This underestimation becomes increasingly large with increasingly smaller sample sizes. To see how this might come about, assume that we are measuring the IQs of randomly selected members of the general population. Suppose the sample has only 5 subjects ($N = 5$). It is unlikely that we will find individuals who are either exceedingly bright or exceedingly retarded in such a small sample. As the sample size is increased, the likelihood of individuals in both extremes being included increases. Therefore, small samples tend not to contain extreme scores, and larger samples more nearly include the entire range of subjects. The extreme condition is reached when the sample size is equal to that of the population, in which case all members of the population are measured and the statistic, based on the sample, is a perfect measure of the population parameter.

The problem is to modify the formula for the SD to correct for this bias. An adequately unbiased value can be obtained with the following formula, in which s is used, in place of σ, to denote the corrected statistic:

$$\sigma \text{ corrected for bias} = s = \sqrt{\frac{\sum (X - \bar{X})^2}{N - 1}} \to \sigma_{\text{pop.}} \qquad (7.4)$$

And for the example:

$$s = \sqrt{\frac{\sum (X - \bar{X})^2}{N - 1}} = \sqrt{\frac{14}{2}} = \sqrt{7} = 2.65 \qquad (7.5)$$

The correction is obtained by replacing (N), as a measure of sample size, with $(N - 1)$. In the calculation we can, to a large extent (but not entirely), eliminate the bias (i.e., the tendency to underestimate) of the SD by dividing the sum of the squared deviations, not by the sample size, but by the *degrees of freedom*.

As with χ^2 for a single sample, where 1 df is lost for the fixed value (the marginal frequency) and $df = C - 1$, here also 1 df is lost for the one fixed value in the calculation (the mean). For example, if three numbers have a mean of 7 and the first two numbers are 4 and 8, then the value of the last number is fixed, since the three numbers must have a sum of 21 (7 × 3). In other words, if the mean value is known, how much additional information can we gain by finding the value of each of the scores? All scores but the last one are informative, so there are $N - 1$ units of information in the group of scores. Degrees of freedom (df), then, is a measure of the available information.

The correction for bias in the calculation of SD is automatically (and appropriately) adjusted to the sample size. With a small sample size (e.g., $N = 3$), dividing the term $\sum (X - \bar{X})^2$ by 2 instead of 3 makes a large difference. With a larger sample size (e.g., $N = 30$), dividing by 29 instead of 30 makes a relatively small difference; and when the sample size gets still larger (e.g., $N = 100$), the difference between the biased and unbiased measure of SD is negligible.

CALCULATION OF THE STANDARD DEVIATION

Although the standard deviation can be calculated from the definitional formula (7.4), this is a cumbersome and awkward procedure. In the text examples the means are small whole numbers, but in real life (i.e., outside the text and outside the classroom), we are likely to find values such as 17.89 and 53.47, and it is very awkward to subtract such a value from each of the scores and square the resulting term. Furthermore, the value for the mean often must be rounded off, so that the deviation scores (i.e., $X - \bar{X}$) will involve a slight rounding error, which will lead to slight inaccuracies. Formulas that are algebraically equivalent to the definitional formula for SD can readily be derived. In these formulas, which are admirably suited for use with an electric calculator, all terms are in the form of raw scores (original scores as opposed to deviation scores). A convenient version of the raw-score formula is:

$$s = \sqrt{\frac{N \sum X^2 - (\sum X)^2}{N(N - 1)}} \tag{7.6}$$

As an example, the raw scores for the original set of quiz scores (Table 7.1) are calculated in Table 7.4. The SD can be calculated using the definitional formula:

$$s = \sqrt{\frac{\sum (X - \bar{X})^2}{N - 1}} = \sqrt{\frac{14}{2}} = \sqrt{7} = 2.65 \tag{7.7}$$

and the raw-score formula:

$$s = \sqrt{\frac{N(\sum X^2) - (\sum X)^2}{N(N - 1)}} = \sqrt{\frac{3(161) - (21)^2}{3(3 - 1)}} = \sqrt{\frac{483 - 441}{3(2)}}$$

$$= \sqrt{\frac{42}{6}} = \sqrt{7} = 2.65 \tag{7.8}$$

TABLE 7.4

X	X^2	$\bar{X}$	$(X - \bar{X})$	$(X - \bar{X})^2$
4	16	7	-3	9
8	64	7	$+1$	1
9	81	7	$+2$	4
$\sum X = 21$	$\sum X^2 = 161$			$\sum (X - \bar{X})^2 = 14$

NORMAL DISTRIBUTION

In Chapter 1 the algebraic expansion of the binomial term $(p + q)^N$ was discussed. Figure 7.6 shows the distribution of outcomes given by the binomial expansion for increasing values of N (i.e., increasing number of tosses) for the coin-tossing example in which $p = q = 1/2$. The distribution

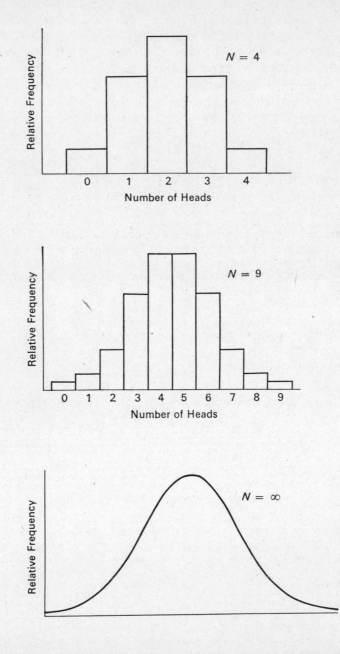

is symmetrical, or equally shaped, on either side of the mean. As N increases, the distribution becomes less steplike. When N is infinitely large, the distribution has a bell shape, which is found in the distributions of many natural phenomena such as height, weight, and IQs. Since the distribution is symmetrical and bell shaped, the mean and median coincide, and most cases fall near the mean with the number of cases dropping sharply with increasing distance from the mean in either direction.

The distribution for $N =$ infinity can be specified mathematically and is called the **normal distribution** (ND). The normal distribution is frequently encountered in the measurement of many variables, but its main importance at this point is that many statistical measures have sampling distributions that are essentially normal in shape. Assigning probability values to statistical tests with normal sampling distributions requires a knowledge of the characteristics of the ND. Furthermore, the statistical procedures covered in the following chapters involve the assumption that the samples are randomly selected from normally distributed populations—an assumption that is difficult to test and, in practice, can be violated somewhat without doing serious damage to the tests. In most cases where the assumption is clearly untenable, we can use a less demanding, if slightly less efficient, test—usually a ranking test. The problem of selecting the appropriate analysis will be discussed in more detail in the following chapters.

If we know the mean and the SD of a distribution and that the distribution is a normal one (i.e., normally distributed), then we have essentially all of the information needed to describe that distribution. For example, consider a distribution of IQs for the total population of young adults. This is a normally distributed characteristic having a mean of 100 (which is an arbitrarily set value in accordance with the definition of IQ) and a standard deviation of 16 (which is determined empirically on the basis of testing a large sample). Thus we have a ND, and we know the mean and SD. Is there anything more we can say about the distribution? Yes. Since the distribution can be defined mathematically, it is possible to determine the proportion of the total area under the curve that lies within any given region. Tables have been devised giving these areas, thereby eliminating the need for calculation. Table D is a typical version of such a table.

Before considering the ND table, we have to return to the standard deviation. The SD is a statistic relating to the degree to which the scores are spread about the mean of the distribution. In the present example the value of the SD $= 16$ IQ points. Using 16 as the value of 1 *standard deviation unit*, we can mark the SD units off on the base line of the distribution, as done in Figure 7.7.

The range of the ND is approximately 3 SD (3σ) units above and below the mean. (Over 99 percent of all cases within the ND fall within this total range of 6σ.) Furthermore, some 2/3 of the cases within the distribution occur within a range of 1σ unit above and below the mean (i.e., $\pm 1\sigma$).

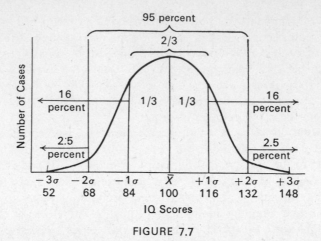

FIGURE 7.7

About 95 percent of the total cases occur within a range of $\pm 2\sigma$ of the mean. Given this information we can answer several specific questions with regard to IQ:

1. What proportion of the subjects has an IQ greater than 116? The first step is to identify the area under the normal curve that we are interested in. An IQ of 116 is 16 points above the mean ($116 - 100 = 16$). In terms of SD units, which in this case $= 16$ IQ points, we are dealing with 16 IQ/16 IQ $= 1$, or 1 SD unit (i.e., 1σ) above the mean. The problem is to find what proportion of the subjects or, in terms of the curve, what proportion of the area under the curve lies beyond the point marked off by an IQ of 116—that is, what proportion lies from this point to the high end of the curve. Looking at Figure 7.7, we can see that 2/3 of the cases lie within $\pm 1\sigma$ of the mean, and 1/3 of the cases fall outside this limit. Therefore, 1/2 of this 1/3, or 1/6 (about 16 percent), of the total cases fall in the portion of the distribution beyond 116.

2. Similarly, what percentage of young adults in the general population has an IQ below 68? Using the same rationale, the deviation in terms of SD units is calculated as follows:

$$\text{deviation from } \bar{X} \text{ in SD units} = \frac{\text{score} - \bar{X}}{\sigma} = \frac{68 - 100}{16}$$

$$= \frac{-32}{16} = -2 \qquad (7.9)$$

Since approximately 95 percent of the cases fall within 2σ units of the mean, 1/2 of the remaining 5 percent, or about 2.5 percent, would fall in the portion of the curve below the value of 68.

Z SCORE, OR STANDARD SCORE

We have been calculating the location of a score in terms of its deviation from the mean in SD units. This score is known as a **Z score**, or **standard score**, and has the following formula:

$$Z = \frac{X - \bar{X}}{\sigma} \quad \text{or} \quad \frac{X - \bar{X}}{s} \tag{7.10}$$

The Z score is a measure of the location of the original score within the distribution. Scores higher than the mean have positive Z scores (e.g., $Z = +1.23$), and scores lower than the mean have negative Z scores (e.g., $Z = -.83$). Z scores provide a convenient way of comparing scores from distributions that have different means or standard deviations. Suppose, for example, student A gets a grade of 80 on a history quiz and student B scores 76 on a biology exam. Who did better? There is no meaningful way to compare the two grades without knowing the distributions of scores for each class. This information is given and the comparison shown in Table 7.5. Student A, even though he had the higher numerical grade, clearly did

TABLE 7.5

Student	Grade	$\bar{X}$	s	Z
A	80	84	8	$\frac{80 - 84}{8} = -.5$
B	76	60	10	$\frac{76 - 60}{10} = +1.6$

worse (relative to the rest of the group) than student B. Z scores can be used to compare scores for any measures for which the distribution is known. On tests such as College Board or Graduate Record examinations, the reported "scores" are modified Z scores, which permit a direct comparison of a student's performance on the verbal and quantitative sections. The use of standard scores in reporting test scores is discussed further in Supplement G.

PROBABILITY AND THE NORMAL DISTRIBUTION

Since Z scores for normal distributions are directly comparable, a single table (Table D) is all that is needed to give the proportional area in a ND beyond any point measured in terms of Z scores.

Instead of thinking in terms of the proportion of the total area lying beyond a given point in the distribution, we can think in terms of the chance,

TABLE 7.6. Table of Probabilities Associated with Values as Extreme as Observed Values of Z in the Normal Distribution

z	.00	.01	.02	.03	.04
.0	.5000	.4960	.4920	.4880	.4840
.1	.4602	.4562	.4522	.4483	.4443
.2	.4207	.4168	.4129	.4090	.4052
.3	.3821	.3783	.3745	.3707	.3669
.4	.3446	.3409	.3372	.3336	.3300
.5	.3085	.3050	.3015	.2981	.2946
.6	.2743	.2709	.2676	.2643	.2611
.7	.2420	.2389	.2358	.2327	.2296
.8	.2119	.2090	.2061	.2033	.2005
.9	.1841	.1814	.1788	.1762	.1736
1.0	.1587	.1562	.1539	.1515	.1492
1.1	.1357	.1335	.1314	.1292	.1271
1.2	.1151	.1131	.1112	.1093	.1075
1.3	.0968	.0951	.0934	.0918	.0901
1.4	.0808	.0793	.0778	.0764	.0749

Z = 1.00

area = p = .1587

or probability, of randomly selecting an individual having certain character-istics. For example, what is the probability of randomly selecting a subject and finding that he has an IQ greater than 116? As we worked it out before, the Z score is 1.0. Table 7.6 gives the appropriate section of Table D and indicates that .1587 of the total area or of the total number of subjects lies in this region of the distribution. In other words, since 15.87 percent of all subjects fall in this region, the chance of randomly selecting an individual who has an IQ over 116 would be 15.87 percent. Similarly, the chances of finding a randomly selected individual with an IQ of less than 68 is 2.28 percent.

In the next few chapters we will work with statistical measures of the differences between what is observed and what would be expected by chance, and these measures will have sampling distributions that are essentially normal. The appropriate probability values will be determined in basically the same way as with Z scores and the normal distribution.

†† COEFFICIENT OF VARIABILITY

Newborn babies vary in size (assume $s = 1.0$ pound and $\bar{X} = 8$ pounds), as do their mothers ($s = 15.0$ and $\bar{X} = 125$). Can we reasonably conclude from the difference in s values that mothers are more variable in weight than their offspring? It would be misleading to compare the s values directly. If the mothers' SD equaled 1.0, the weights would be remarkably the same for all mothers. On the other hand, since the range of scores is approximately ± 3 SD about the mean, a SD of 15 for infants is clearly impossible. Therefore, the variability of mothers should not equal that of infants, in view of the large difference between the average weights of the two groups. What happens if we take the mean value into account in assessing the variability? This is conveniently accomplished by finding the ratio of the standard deviation to the mean, or the **coefficient of variability** (CV):

$$CV = \frac{\text{standard deviation}}{\text{mean}} = \frac{s}{\overline{X}} \tag{7.11}$$

which is sometimes expressed as a percentage by:

$$CV = \frac{s(100)}{\overline{X}} \tag{7.12}$$

For the example, the calculations would be:

$$CV \text{ for infants} = \frac{1 \text{ pound}}{8 \text{ pounds}} = .125 = 12.5 \text{ percent} \tag{7.13}$$

$$CV \text{ for mothers} = \frac{15 \text{ pounds}}{125 \text{ pounds}} = .120 = 12.0 \text{ percent} \tag{7.14}$$

Since the CV values for both groups are essentially the same, infants and mothers appear to be equally variable (in weight). The constant CV value indicates that the increase in the variability is in proportion to the mean.

Situations in which there is a positive relationship between the mean $(\overline{X})$ and the standard deviation (i.e., s) are fairly common, particularly when the overall level of scores is low and therefore the opportunity to vary reduced—as in the case of infant weights. Data that show a relatively constant CV for the different groups are often given a logarithmic transformation to equalize the variances of the groups prior to statistical analysis (see Supplement B on transformations).

Note that in equations (7.13) and (7.14) the unit of measurement (pounds) cancels out. Since the CV value is a unit-free ratio, it can be used to compare different types of measures. To determine whether the infants' weight is more variable than their length, for instance, we can compare the CV for weight to the CV for length. The CV measure cannot be used, however, to compare two variables when the zero point for either one is an arbitrary, rather than an absolute, value. The Fahrenheit and Centigrade temperature scales and certain psychological scales have arbitrary zero points, whereas most physical measurements (e.g., height and weight) have absolute zero points.

SUMMARY

1. *Categorical data* provide information about only the category in which the measurement falls. Scores are obtained when the categories are limited to the degree or level of a given characteristic. *Ordinal data* provide information concerning the order, or rank, of the scores, and *interval data* provide information concerning the order and the spacing, or intervals, between adjacent scores.

2. A measure of *central tendency* (an "average") is a single score that can be taken as representative of a group of scores.

3. The *mean* is the sum of scores divided by the number of scores; that is, $\bar{X} = \sum X/N$. It is the most frequently used measure of central tendency and is most suitable for symmetrically distributed data.

4. The *median* is the "middle" score and is used for ordinal data. For interval data that are skewed, the median is often preferable to the mean.

5. The mean has the property that the sum of the squared deviations from the mean of each score in the distribution is less than the sum of squared deviations about any other point. Therefore, the mean is determined by the *method of least squares*.

6. The *mode* is the score that occurs more often than any other. It is the least satisfactory and least frequently used measure of central tendency.

7. A set of scores that has the same distribution on both sides of the mean is a symmetrical, or *nonskewed*, distribution. If the scores on one side of the mean are stretched out or compressed compared to the other side, the distribution is *skewed*.

8. The *standard deviation* (SD) is the square root of the mean value of the squared deviations about the mean and is the most frequently used *measure of dispersion* or spread of scores.

9. The SD of a sample calculated with the definitional formula for σ tends to be *biased* in that it *underestimates* the SD of the population from which the sample was obtained. An *unbiased estimate* of the population SD (s) can be obtained by dividing $\sum (X - \bar{X})^2$ by $(N - 1)$ instead of by (N).

10. The term $(N - 1)$ in the calculation of s is the *degrees of freedom* in the sample. In this case $(N - 1)$ equals the number of scores conveying information if the mean value is fixed or known.

11. The *normal distribution* (ND) is a bell-shaped distribution. Many natural phenomena and chance events (including the sampling distributions of many statistics) are normally distributed.

12. The *Z score* is a score expressed as a deviation from the mean in standard deviation units; that is, $Z = (X - \bar{X})/\sigma$, or $(X - \bar{X})/s$.

13. If a given event is distributed normally, the proportion of the total area under the curve falling beyond a given score (expressed as a Z score) can be readily determined using Table D. This area is equal to the proportion of cases and indicates the *probability* of obtaining scores as great or greater than the given score.

†† The *coefficient of variability* (cv) is the ratio of the sd to the mean $(s/\bar{X})$ and is used to compare variability in situations where the variability increases with increases in the mean.

chapter eight

Chapter 6 dealt with the rank-order correlation coefficient—a measure of the relationship between two variables for data expressed in ranks. The rank-order correlation coefficient is a measure of the degree to which an individual's relative standing within a group, as indicated by his ranking, is similar or dissimilar for two variables. In other words, it measures the degree to which a ranking, or lining up, of subjects with regard to one variable (e.g., height) corresponds to the ranking of the subjects with regard to another variable (e.g., weight). The rank-order coefficient indicates how well we can estimate or predict an individual's ranking in terms of weight, for example, if we know his ranking in terms of height.

PEARSON CORRELATION COEFFICIENT

Let us now consider an equivalent statistic for interval data. As with the rank-order correlation, we need an appropriate measure of each subject's relative standing within the sample for each of the variables measured. Since the scores are not ranked, we need a measure of relative position in terms of the numerical values of the data. An obvious measure is the score relative to the mean of the sample (i.e., $X - \bar{X}$), but this is not satisfactory, since the numerical value of the difference depends upon the particular unit of measurement employed (e.g., 2 inches or 50 cm.). We need a measure of relative standing that is unaffected by the original unit of measure. The appropriate measure is the deviation score ($X - \bar{X}$) expressed in standard deviation units—the familiar Z score, or standard score:

$$Z = \frac{X - \bar{X}}{s} \tag{8.1}$$

109

The relationship, then, is measured in terms of the correspondence of the Z scores of each subject for the two variables in the same fashion that the rank-order coefficient compares the relative rankings of the two measures for each subject. The correlation measure for interval data is known as the **Pearson correlation coefficient** (r), which has the following definitional formula in terms of the average products of the Z scores for each subject:

$$r = \frac{\sum (Z_X Z_Y)}{N} \qquad (8.2)$$

where $\quad Z_X = \dfrac{X - \bar{X}}{s_X}$

$\quad Z_X Z_Y$ = product of Z scores for each subject
$\quad N$ = number of subjects

To understand formula 8.2, consider a situation in which the two variables are fairly highly related (such as height and weight). Individuals who are above the mean on one variable ($Z_H = +$) will tend to also be above the mean on the other variable ($Z_W = +$), and $(+) \times (+) = (+)$. Similarly, individuals below the mean on one variable $(-)$ will tend to be below the mean $(-)$ on the other variable, and $(-) \times (-) = (+)$. When the two sets of relative Z scores correspond, the value for the final statistic (r) will be the maximum positive value. In a situation in which the relative positions of one variable are inversely related to the other variable (e.g., height of basketball players and head clearance through doorways), subjects above the mean $(+)$ on one variable are correspondingly below the mean $(-)$ on the other variable, and $(+) \times (-) = (-)$. Similarly, individuals below the mean $(-)$ on the first variable are above the mean $(+)$ on the second variable, and $(-) \times (+) = (-)$. Therefore, the value for r will be the largest possible negative value. Finally, when the subjects' relative standings on the two variables are essentially unrelated, many different combinations of $(+)$ and $(-)$ will occur, and the final positive and negative products will cancel to give $r = 0$. As with the rank-order coefficient the degree of relationship is measured by the value of the correlation, which will be from .00 to 1.0, and the direction of the relationship is indicated by the sign of the correlation.

We would also like to know if the obtained correlation coefficient is likely to have occurred in the sample by chance if there is no correlation in the population. A sampling distribution could be determined for r in the same manner as for r_s. The critical values for the significance of r are given in Table N. Note that the degrees of freedom for the correlation is $(N - 2)$. The mean values for each of the two measures are the two fixed values that reduce df. The probability values are similar, but not identical, to those for r_s. The probability value for r can also be obtained with a t test (discussed in Chapter 9) when N is large and r is small:

$$t = \frac{r}{\sigma_r} = \frac{r\sqrt{N-2}}{\sqrt{1-r^2}} = \sqrt{\frac{r^2(df)}{1-r^2}} \tag{8.3}$$

In general, Table N is easier to use and more accurate than formula 8.3, particularly when N is small and r is large.

When we present a correlation coefficient (whether the Pearson or rank-order statistic), it is necessary to provide information concerning: (1) the degree of relationship (r), (2) the direction of the relationship $(+$ or $-)$, and (3) the likelihood that the relationship might be due to chance (p).

Calculating the value for r with the definitional formula is very cumbersome, since it involves obtaining a mean for each group, the Z score for each measurement, the product of the Z scores for each individual, and the mean of these products. Aside from the many steps, appreciable rounding errors can enter into the calculation when the mean is not a whole number. In the same way that the value for s can be obtained with a raw-score formula, the formula for r can also be expressed in terms of the original scores. A formula for r, which is well suited for electric or electronic calculators, is:

$$r = \frac{N \sum XY - (\sum X)(\sum Y)}{\sqrt{N \sum X^2 - (\sum X)^2} \sqrt{N \sum Y^2 - (\sum Y)^2}} \tag{8.4}$$

when X = measurements on one variable
$\quad\quad Y$ = measurements on the other variable

Note that the denominator contains the major part of the raw-score formula for the standard deviation (s):

$$\sqrt{N \sum X^2 - (\sum X)^2}$$

when
$$s = \sqrt{\frac{N \sum X^2 - (\sum X)^2}{N(N-1)}} \tag{8.5}$$

With the background material covered, we can now see how the correlation measure works in practice. At the local zoo a very famous and popular attraction features the keepers in the fish house feeding the piranhas and sharks by hand. Part of the attraction is that every once in a while one of the beasts is just a little bit extra fast at the same moment that one of the keepers is just a little bit extra slow. You are (for some reason) interested in the relationship between the duration of employment as a keeper in the fish house (X) with the number of fingers remaining on the hands of the keeper (Y). You count each joint as equal to 1/3 finger, except for the thumb, which is scored as 1/2 finger per joint. This gives a maximum possible total of

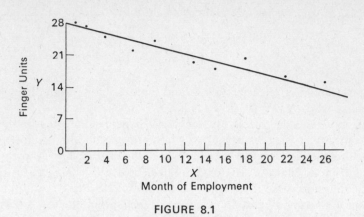

FIGURE 8.1

$(4 \times 3 + 2)$ joints $\times$ 2 hands = 28 total finger units. For the current group
of keepers you obtain the data shown in Table 8.1. A scattergram of these
data (in terms of scores—not ranks as used for the rank-order correlation)
provides the picture shown in Figure 8.1. The slope from the upper left down
to the lower right indicates that the relationship is negative, and the spread of
the data points about the straight line (which has been fit by eye) shows that
the relationship is fairly high. As with the rank-order correlation, the Pearson
coefficient is a direct function of the degree to which the data points on the
scattergram can be approximated by, or fit with, a straight line. The scores
for both the number of finger units and the length of time employed are fairly
normally distributed, and the mean and SD are suitable measures. The
Pearson coefficient, therefore, is more appropriate than the rank-order
correlation for these data.

TABLE 8.1

Subject	X (months)	X^2	Y (finger units)	Y^2	XY
a	1	1	28	784	28
b	2	4	27	729	54
c	4	16	25	625	100
d	7	—	22	—	—
e	9	—	24	—	—
f	13	—	19	—	—
g	15	—	18	—	—
h	18	—	20	—	—
i	22	—	16	—	—
j	26	—	15	—	—
$N = 10$	$\sum X = 117$	$\sum X^2 = 2{,}029$	$\sum Y = 214$	$\sum Y^2 = 4{,}764$	$\sum (XY) = 2{,}171$

The correlation coefficient is calculated as follows:

$$r = \frac{10(2,171) - (117)(214)}{\sqrt{10(2,029) - (117)^2}\ \sqrt{10(4,764) - (214)^2}} = -.9539$$

$$p < .001 \tag{8.6}$$

$$s_X = \sqrt{\frac{10(2,029) - (117)^2}{10(9)}} = 8.5641 \tag{8.7}$$

$$s_Y = \sqrt{\frac{10(4,764) - (214)^2}{10(9)}} = 4.5265 \tag{8.8}$$

There is a significant ($p < .001$), very high, negative correlation ($r = -.95$) between length of time employed and finger units; the longer the men worked, the fewer fingers they had remaining.

The two correlation coefficients (r and r_s) will generally be close in value for the same data and will tend to differ as the data diverge from being normally distributed. The rank-order formula can be algebraically derived from the Pearson formula. A Pearson correlation coefficient based on *ranks* of the scores will always produce the same value as the calculations for r_s. The rank-order formula, therefore, permits simplified calculations for ranked data, but considerable information beyond the correlation coefficient is available when dealing with the scores as interval data.

BEST-FIT, OR REGRESSION, LINES

Assume (as with the rank-order correlation) that the relationship between the two variables is linear; that is, it can be depicted by a straight line. The failure of the scattergram points to all fall on a line can be attributed to chance, or random, effects. The problem then becomes one of "uncovering" or calculating the underlying straight line from the observed data points. We have previously dealt with the problem of obtaining a single score that represents a sample—that is, the mean. The mean is fitted by the method of least squares; that is, the sum of the squared deviations from the mean [i.e., $\sum (X - \bar{X})^2$] is less than the sum of the squared deviations from any other point in the distribution. We can specify that the line used to describe the scattergram also be determined by the method of least squares, that is, that the sum of the squared deviations from the line be less than the sum of the squared deviations from any other line. Such a line is called a **best-fit line**, or **regression line**. The mean is the best *estimate* of the scores in the group. Similarly, the regression line represents the best estimate of one variable for each value of the other variable. The formula for the best-fit line is reasonably simple and direct, though not quite as easy as $\sum X/N$. It is given in terms of predicting the Y variable (Y') for a given value of the X variable or for predicting X (i.e., X') from the Y variable:

$$Y' = r\left(\frac{s_Y}{s_X}\right)(X - \bar{X}) + \bar{Y} = aX + b \qquad (8.9)$$

$$X' = r\left(\frac{s_X}{s_Y}\right)(Y - \bar{Y}) + \bar{X} = aY + b \qquad (8.10)$$

where Y' = the value to be estimated or predicted

   X = a given value for which the associated Y value (Y') is being calculated

   a = the slope of the line

   b = the intercept of the line

Once the calculations for the formula have been done, any two values of X can be substituted into the formula to permit graphing of the best-fit line. We can now return to the problem of the zoo keepers and the calculations from equations (8.6), (8.7), and (8.8):

$$Y' = -.9539\left(\frac{4.5265}{8.5641}\right)(X - 11.7) + 21.4$$

$$= (-.5042)X - (-.5042)11.7 + 21.4$$

$$= -.5042X + 27.3 \qquad (8.11)$$

Graphing the best-fit line on the scatter diagram gives the picture in Figure 8.2.

In the same way that the mean summarizes a group of scores, the best-fit line summarizes the nature of the relationship between the two sets of measurements. In the present example, it shows that the average finger loss per month is appreciable. In other words, the **slope** of the best-fit line, which is the term a in the formula, is the measure of the degree to which the Y variable changes with every unit change in the X variable. On the average .5042 fingers were lost during each month of employment. Therefore, in the example data, not only is there a significant ($p < .01$), high, negative correlation ($r = -.95$) between fingers remaining and duration of employment, but the average rate of finger loss (.5042 finger units per month) is high (marvelously so in terms of zoo attendance).

In general, it is valuable to refer to the slope of the best-fit line in addition to the degree, direction, and probability value for the correlation coefficient. In many situations the slope of the best-fit, or regression, line is the most valuable and informative measure of the data. For example, assume we find a high correlation (e.g., $r = +.80, p < .001$) between the amount of fertilizer on a cornfield (or number of weeks of a special reading readiness program for schoolchildren) with the yield of corn (or the reading readiness grade equivalent). The correlation looks very impressive; in fact, we might be tempted to report our results as an important finding. However, consider how different the interpretation might be if the slope indicated that the value of the

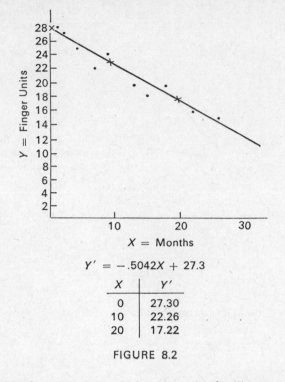

$$Y' = -.5042X + 27.3$$

X	Y'
0	27.30
10	22.26
20	17.22

FIGURE 8.2

additional yield of corn was worth *less* than the fertilizer used or that the improvement in reading readiness was less than would be accomplished with the equivalent tutoring time spent in the regular school program. The correlation coefficient tells how *accurately* one measure can be predicted from the other, but the slope of the best-fit line tells how *much* one variable changes with changes in the other variable.

The best-fit line was described previously as being unique; no other line can be drawn that has its properties. Yet, strange as it may seem, *two* unique best-fit lines can be drawn to a given scattergram if we use formulas (8.9) and (8.10). One line is used to predict the *Y* variable on the basis of the *X* variable (i.e., the number of fingers the zoo keeper will probably have left after he has been working three months), and the other best-fit line is used to predict the *X* variable from the *Y* variable (the length of time he has probably been working there if he has eight fingers left). Another example may help clarify this point: If you had to guess the age of a woman with 8 children, you might guess reasonably that she is 40 years old. If, however, you had to guess the number of children that a 40-year-old woman has, the best guess would certainly not be 8, but rather in the range of 2 or 3. The relationship between these two predictions is shown in Figure 8.3. One of the best-fit lines minimizes the deviation of points from the line when measured in a vertical direction (predicting *Y* from *X*), and the other line minimizes the

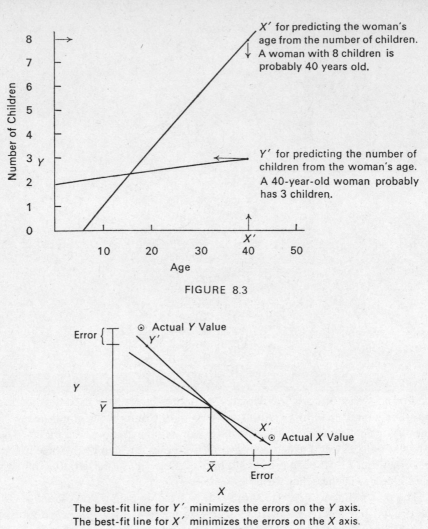

FIGURE 8.3

The best-fit line for Y′ minimizes the errors on the Y axis.
The best-fit line for X′ minimizes the errors on the X axis.

FIGURE 8.4

errors measured in a horizontal direction (predicting X from Y), as shown in Figure 8.4.

The two best-fit lines will always cross at the means of the two distributions. If $\bar{X}$ is substituted for the X term in formula (8.9), the value for Y' will always be equal to $\bar{Y}$. When $r = 1.0$, the two best-fit lines are identical and superimposed, and all data points fall on that line.

ERRORS OF ESTIMATE

Look again at the best-fit line for predicting Y from X for the zoo-keeper example, as shown in Figure 8.5. The smooth curves sketched in the figure

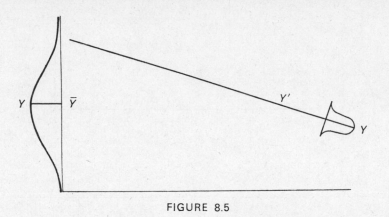

FIGURE 8.5

show the distribution of the scores for finger units (Y) and the distribution of the scores around the best-fit line (Y'). These distributions are only suggested by the few measurements used in the example but are pictured here as if the sample were large. Since the regression line provides the values for estimating or predicting Y (i.e., Y'), the failure of the actual scores to fall on the line means that the predictions are in error by the degree to which each data point deviates from the regression line.

Therefore we need a measure of the accuracy of prediction when using the regression line. The basic measure of error is in terms of the deviation of a score from the regression line ($Y - Y'$). Summing the $Y - Y'$ values for all the points in the scattergram would not be satisfactory. Some of the scores are above the regression line, giving a positive value to the deviation, while others are below, giving a negative value to the deviation. The + and − deviations will (of course) cancel out perfectly so that $\sum (Y - Y')$ will equal 0. This problem is solved by squaring the values of the deviations and then adding these terms, that is, $\sum (Y - Y')^2$. But not all samples contain the same number of subjects, so the number of deviation scores must be taken into account, which can be done by dividing by ($N - 1$) (or df), giving the following formula:

$$\text{error} = \frac{\sum (Y - Y')^2}{N - 1} = s^2 \tag{8.12}$$

The measure of accuracy of estimation, or prediction, associated with a correlation coefficient, then, is given by the variance (standard deviation squared) of the distribution of scores about the best-fit line.

Now it is worthwhile to consider a different formula for the correlation coefficient, one that is algebraically derived from the definitional formula:

$$r^2 = 1 - \frac{s^2_{Y-Y'}}{s^2_{Y-\bar{Y}}} \tag{8.13}$$

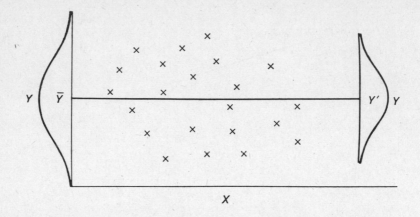

To simplify matters, we will assume that the value for the term $s^2_{Y-\bar{Y}} = 1$. When the correlation is perfect, all the points will fall on the best-fit line, and our ability to predict is perfect. In this case there is no spread of points about the best-fit line, and $s^2_{Y-Y'} = 0$; so $r^2 = 1 - 0 = 1$, and $r = 1$. At the other extreme, when there is no relationship between the two variables, the scattergram will look like that shown in Figure 8.6. Since Y' (the predicted value of Y) $= \bar{Y}$ for all values of X, the X variable provides no help in estimation, and $\bar{Y}$—the best-fit point—is the best guess for Y' for every value of X. Under these conditions $s^2_{Y-Y'} = s^2_{Y-\bar{Y}}$ and $s^2_{Y-Y'}/s^2_{Y-\bar{Y}} = 1$, since the spread of scores about the best-fit line ($Y - Y'$) is identical to the

TABLE 8.2

r	r^2	$s^2_{Y-Y'}/s^2_{Y-\bar{Y}}$	$1 - s^2_{Y-Y'}/s^2_{Y-\bar{Y}} =$ Reduction in $s^2_{Y-Y'}$ Relative to $s^2_{Y-\bar{Y}}$
.0	.0	1.00	.00
.1	.01	.99	.01
.2	.04	.96	.04
.3	.09	.91	.09
.4	.16	.84	.16
.5	.25	.75	.25
.6	.36	.64	.36
.7	.49	.51	.49
.8	.64	.36	.64
.9	.81	.19	.81

spread of scores about the mean of the original distribution ($Y - \bar{Y}$); and $r^2 = 1 - 1 = 0$ and $r = 0$. Table 8.2 provides the $s^2_{Y-Y'}/s^2_{Y-\bar{Y}}$ ratio for a range of values of r.

The square of the correlation coefficient (r^2), then, is the proportion by which the variance of scores about the best-fit line ($s_{Y'}^2$) is *smaller* than the spread of scores about the mean of the original distribution (s_Y^2). In other words, r^2 is a measure of the degree to which knowledge of the second variable (e.g., height) permits an *improvement* in predicting the values for the first variable (e.g., weight). We can further see from Table 8.2 that a correlation coefficient of $r = .5$ is not halfway between a zero and a perfect correlation, since it reduces the variance of the errors by only 25 percent. A correlation of $r = .7$ is needed to halve the variance of the errors, and a correlation coefficient of $r = .1$ represents a negligible improvement in accuracy (1 percent).

A frequently made error is the interpretation of the correlation coefficient primarily in terms of the probability value. There are articles in scientific journals that proudly display "significant correlations," but the authors manage to ignore the fact that the size of correlation is on the order of .15 or .20. In these cases, the technical term "significant," which refers to the likelihood of the observed outcome occurring by chance, is confused with the common meaning of "significant" as appreciable, large, or important. In the same way that we distinguish between the significance of the differences between two groups (measured by p) and the size of the difference (measured by r_m), when dealing with correlation we distinguish between the probability value for that relationship (measured by p) and the degree of relationship (measured by r). We also have to be careful, when interpreting the relationship between two variables, to look not only at the degree of relationship (r) but also at the manner in which one variable changes as the other variable changes (indicated by the slope of the best-fit line).

CORRELATION AND CAUSALITY

Another problem often arises in the interpretation of relationships between two variables. Imagine that we find a high, negative, and significant correlation between the use of vitamin supplements in family diets and the incidence of tooth decay (e.g., $r = -.6$, $p < .001$). It appears that as more vitamin supplements are consumed, the number of decayed teeth decreases. The regression line supports the conclusion that the effect is fairly large. From these data could we reasonably conclude that vitamin supplements are an aid in reducing tooth decay? It certainly is very tempting to do so, but assume that, in addition to measuring vitamin supplements, we also measure the number of magazines read by each family and find a slightly higher correlation between magazine readership and tooth decay. Could we then assume that, somehow, magazine reading promotes dental health? Actually, we can come to no conclusion concerning either vitamins or magazines, since the presence of the relationship (as measured by the correlation coefficient) tells *nothing* about the causal relationship between the two variables.

Formula (8.4) for calculating r consists of such terms as $\sum X$ and $\sum XY$. Nowhere in the formula is there any reference to causality. The formula is inherently stupid (and this would still be the case even if the calculations were done with the most modern computer) in that it can derive no more from the data than is put into the calculations. A correlation may mean that there is a causal relationship, but proof of such a relationship must depend upon information in addition to that which goes into the calculation of the correlation itself. The correlation values in the above example are reasonable ones, since the use of vitamin supplements and magazine readership tend to increase with increased socioeconomic levels and so does access to appropriate dental and medical care. Therefore, the relationship among the variables of magazine readership, vitamin consumption, and tooth decay is probably due to the fact that all three of these variables reflect the general level of prosperity of the family. We could expect the amount of money spent on vacations and the cost of the family automobile(s) to be similarly correlated with the other three variables.

The temptation to attribute causality on the basis of a correlation coefficient is particularly strong when one variable precedes the other variable in time. We could feel this temptation when thinking over the vitamin supplement and tooth decay situation. A common (but less important) example of this type of error is the statement that "the rain really cooled down the weather, didn't it?" Actually, the rain does not cause the air to become cool—though the rain comes before the change in temperature; rather, it is the edge of the mass of cool air (the cold front) that triggers the rain. The cool air causes the rain that apparently precedes it.

In these two examples it is relatively easy to see the mistake, but in practice it is just as easy to be misled by data suggesting a causal relationship that is not really there. Statements of causality must be based on knowledge of the situation in addition to the information used for the calculation of the correlation coefficient. Correlations frequently occur when the two variables being measured are both influenced by a (hidden) third variable. When one of the measured variables precedes the other in time, we have to be doubly cautious in deciding whether the relationship is also a causal one. Experimental studies in which the independent variable is manipulated and precedes the dependent variable can be used to establish whether or not a given relationship is also a causal one.

PREDICTION

Much of the discussion about correlation has been concerned with the problems of estimation, or prediction. This might seem unnecessary. If, for example, we have the heights and weights of a group of subjects, there really is no need to "guess" the weight of a given subject on the basis of his height; we already know what his weight is. This would be true if we were only interested in that specific group of subjects, but usually we are concerned with the information the sample can provide about the population from which it was drawn. With the information afforded by the sample, we can deal with any possible case. For example, we could determine the best estimate of the weight of a subject who is 6 feet, $3\frac{1}{2}$ inches tall, even though none of the subjects in the sample happen to have that exact height.

Prediction is particularly important when applicants for college admission are considered. On the basis of an applicant's high school record or the score on his college entrance examination, a reasonable prediction of what his performance will be in college can be obtained by referring his score to the regression line for the previous year's entering class. The admissions committee predicts that, if a student with this background had been admitted last year, he would be performing at such and such a level this year. Minimum standards for admission are often established in this way, and decisions concerning an individual student are influenced by such predictions. These predictions are not infallible, since they are subject to a known degree of error (given by $s_{Y-Y'}$) and are based on the assumption that the students from prior years are a satisfactory approximation of the entering class.

The following example further illustrates the use of regression lines for prediction. In a study of the metabolic rate of lizards at various temperatures, it was necessary to determine the body volume of each subject, which was accomplished by submerging each animal in water and measuring the volume of the displaced water. This process was disturbing for both the subjects and the experimenter, and only the initial batch of lizards was measured in this fashion. Each subject was also weighed, and the regression line based on the volume and weight of each subject was calculated. Subsequent subjects were merely weighed and their volume estimated with the regression equation. The correlation between weight and volume ($r = .91$) indicated that the estimates would be adequate for the study. The correlation coefficient (r) was used here as a measure of the accuracy of the prediction, but the actual prediction was based on the regression equation.

There are extensions of the correlation procedure that permit the use of two or more measures to give a more accurate prediction. This procedure, called multiple correlation, is more complex than simple correlation and is briefly discussed at the end of this chapter.

In most of the examples used in discussing correlation, the two variables have been qualitatively different (e.g., months and fingers, fertilizer and corn). A problem sometimes arises when the two variables are based on the same measure. Consider, for example, a grade school enrichment program where we measure the reading readiness of each child before and after the program. What is the appropriate test for these data? This depends on the specific question being asked of the data. If we wish to ask whether the reading readiness of the children is greater after the program than before, the appropriate test is clearly a t test for related samples. If we want to find out whether the better students are still the better ones and the poorer students still the poorer ones at the end of the program, or in other words, whether the relative standing of the students within the group is the same before and after (regardless of the overall level of performance), then a correlation is the appropriate measure. Note that if the program were infinitely successful, then all the children, regardless of their initial score, would end up with the maximum reading-readiness score, so that the before–after correlation would be effectively 0. A program that had a negligible effect upon the children would result in a near perfect correlation. The interpretation of correlation data in this type of situation requires great care.

In another situation we measure the heights of fathers and their ten-year-old sons. The most obvious question is whether or not tall fathers have tall sons and short fathers have short sons. This would be appropriately answered by a correlation coefficient. But we can also ask whether fathers tend to be the same height as their ten-year-old sons. The appropriate test here is the *t* test for related samples (no pun intended), and we would expect to find that the fathers are appreciably and significantly taller than their young sons.

All combinations of correlations and differences are possible: We might have correlations and differences (e.g., the heights of fathers and their sons), differences but no correlations (e.g., reading readiness after a perfect enrichment program), and correlations with no differences (weights of women when they are having happy thoughts and their weights when they are having unhappy thoughts).

†† PARTIAL-CORRELATION COEFFICIENT

Under certain circumstances the correlation between two variables (e.g., height and weight of children or mechanical ability and grades in physics) can be difficult to interpret, since each of the variables in the pair is known to be influenced by, or correlated with, a third variable (e.g., age or general intelligence). If the correlation between each of the original variables and the third variable is known (and the relationships are appropriately linear), then the influence of the third variable can be eliminated from or partialed out of the correlation between the original two variables. In other words, the correlation between children's height and weight can be determined as if all the children were the same age—the age variable no longer being a factor. The **partial-correlation coefficient** ($r_{12\cdot3}$) for the correlation of variables 1 and 2 with variable 3 partialed out is a function of all of the intercorrelations (r_{12}, r_{13}, r_{23}):

$$r_{12\cdot3} = \frac{r_{12} - r_{13}r_{23}}{\sqrt{(1 - r_{13}^2)(1 - r_{23}^2)}} \qquad (8.14)$$

The significance of the partial-correlation coefficient can be tested by using the t distribution to estimate the sampling distribution of $r_{12 \cdot 3}$ with $df = N - 3$, as follows:

$$t = \frac{r_{12 \cdot 3}}{\sqrt{\dfrac{1 - r_{12 \cdot 3}^2}{N - 3}}} \tag{8.15}$$

For example, the three variables, height (H), weight (W), and age (A), of 30 children yield the following correlations:

$$r_{HW} = r_{12} = .90 \tag{8.16}$$

$$r_{HA} = r_{13} = .70 \tag{8.17}$$

$$r_{WA} = r_{23} = .50 \tag{8.18}$$

which lead to the following calculations:

$$
\begin{aligned}
r_{12 \cdot 3} &= \frac{.90 - (.70)(.50)}{\sqrt{(1 - (.70)^2)(1 - (.50)^2)}} \\[2mm]
&= \frac{.90 - .35}{\sqrt{(1 - .49)(1 - .25)}} \\[2mm]
&= \frac{.48}{\sqrt{(.51)(.75)}} = .78
\end{aligned}
\tag{8.19}
$$

With $N = 30$,

$$
\begin{aligned}
t &= \frac{.78}{\sqrt{\dfrac{1 - (.78)^2}{27}}} \\[2mm]
&- \frac{.78}{.12} - 6.48 \\[2mm]
& \qquad p < .001
\end{aligned}
\tag{8.20}
$$

The example provides a fairly high ($r_{12 \cdot 3} = .78$) and significant ($p < .001$) correlation between height and weight when age is held constant; the taller children tend to be heavier.

As we can see from formula (8.14), if the third variable is unrelated to either of the other two variables (i.e., $r_{13} = 0 = r_{23}$), the value of $r_{12 \cdot 3}$ will equal r_{12}. If all the intercorrelations are positive, then the value of $r_{12 \cdot 3}$ will be less than r_{12} to the degree that the third variable is responsible for the original correlation. This procedure can be extended to partial out, or eliminate, the effects of additional variables.

†† MULTIPLE CORRELATIONS

One of the uses of correlation measures and regression measures is the prediction of an individual's score on one variable on the basis of his score on another variable. In many situations, such as college admissions, it is necessary to make the best possible prediction of performance on the basis of several measures of the individual, such as high school grades and scores on college-entrance exams. Under these conditions the original correlation between two variables and a regression line can be extended to correlations of a given variable with a combination of several other variables, and the regression line is replaced by a regression plane (which adds another dimension to the regression line). This topic is briefly introduced here, a more detailed discussion being beyond the scope of the text.

With only three variables the **multiple-correlation coefficient** ($R_{1 \cdot 23}$) is a function of the intercorrelations of the three variables as follows:

$$R_{1 \cdot 23} = \sqrt{\frac{r_{12}^2 + r_{13}^2 - 2r_{12}r_{13}r_{23}}{1 - r_{23}^2}} \qquad (8.21)$$

The correlation between variable 1 and variables 2 and 3 is greatest when r_{12} and r_{13} are high and r_{23} is low. In other words, the most information about variable 1 can be obtained when variables 2 and 3 provide nonredundant information about the first variable.

The interpretation of $R_{1 \cdot 23}$ is essentially the same as for the simple-correlation coefficient r_{12}, in that the value r^2 is a measure of the degree to which the variance of the predicted scores is less than the variance of the original scores. Similarly, for the three-variable problem, the regression equation for the best prediction of variable 1 from variables 2 and 3 is most easily handled in terms of the standard scores, $(X - \overline{X})/s$. The regression equation for $R_{1 \cdot 23}$ is a function of the intercorrelations of the three variables and has the following form:

$$\frac{x_1}{s_1} = \left(\frac{r_{12} - r_{13}r_{23}}{1 - r_{23}^2} \right) \frac{x_2}{s_2} + \left(\frac{r_{13} - r_{12}r_{23}}{1 - r_{23}^2} \right) \frac{x_3}{s_3} \qquad (8.22)$$

where $x_1 = X_1 - \overline{X}_1$

Knowing both the applicant's high school grades and his college-entrance exam scores, then, we can arrive at a better prediction of his college performance than we could on the basis of either of the measures alone.

The multiple-correlation technique can be extended to more than three variables, but the calculations of the correlation coefficient and of the regression equation become markedly more complex.

PEARSON PRODUCT MOMENT CORRELATION

Formulas

Definitional: $r = \dfrac{\sum (Z_X Z_Y)}{N}$

Raw score:

$$r = \frac{N \sum XY - (\sum X)(\sum Y)}{\sqrt{N \sum X^2 - (\sum X)^2} \times \sqrt{N \sum Y^2 - (\sum Y)^2}}$$

Best-fit line: $Y' = r\left(\dfrac{s_Y}{s_X}\right)(X - \bar{X}) + \bar{Y}$

Table N

provides p values, based on r and N, for the r value being obtained with a sample when there is no correlation in the population.

Restrictions

The relationship between the two variables should be essentially linear for this measure to be meaningful. The two groups of scores should not be skewed but should be normally distributed, and the mean and standard deviation should be suitable measures.

Procedure

1. Add the scores across subjects for each measure to obtain $\sum X$ and $\sum Y$.
2. Square each score and add across subjects to obtain $\sum (X^2)$ and $\sum (Y^2)$.
3. For each subject multiply the X and Y scores and add these products across subjects to obtain $\sum (XY)$.
   Note: with many calculators the values of $\sum X$, $\sum Y$, $\sum (X^2)$, $\sum (Y^2)$, and $\sum (XY)$ can be obtained simultaneously, requiring each X and Y value to be entered only once.
4. Calculate r by substituting the calculated values in the raw-score formula. Note that $N =$ number of subjects.
5. Use Table N to determine the p value based on r and N (where $df = N - 2$).

6. Calculate the regression line for the dependent variable (X' or Y') if information about the rate of change in X with change in Y is desired.
7. Report the direction, size, and probability value of the correlation coefficient as well as the slope of the regression line when available.
8. A scattergram can be used both to determine whether the relationship is linear and to aid interpretation.

Example

Scores on an "interest in science test" (ST) administered to incoming freshmen are compared to their average numerical grades in science courses. Results for a representative sample are:

Subject, $N = 6$	ST (X)	X^2	Grades (Y)	Y^2	XY
a	2	4	68	4,624	138
b	3	9	60	3,600	180
c	5	25	75	5,625	375
d	6	36	83	6,889	258
e	7	49	94	8,836	658
f	9	81	90	8,110	810
	$\Sigma X = 32$	$\Sigma X^2 = 204$	$\Sigma Y = 470$	$\Sigma Y^2 = 37,674$	$\Sigma XY = 2,657$

$$\bar{X} = 5.3333 \qquad \bar{Y} = 78.3333$$

$$r = \frac{6(2,657) - (32)(470)}{\sqrt{6(204) - (32)^2}\ \sqrt{6(37,674) - (470)^2}}$$

$$= \frac{902}{(\sqrt{200})(\sqrt{5,144})}$$

$$= \frac{902}{1,014.2977} = .889 = .89$$

$$s_X = \sqrt{\frac{200}{N(N-1)}} = 2.5819$$

$$s_Y = \sqrt{\frac{5{,}144}{N(N-1)}} = 13.0945$$

From Table N $p < .02$

$$Y' = .89\left(\frac{13.0945}{2.5819}\right)(X - 5.3333) + 78.3333$$

with $\left[.89\left(\dfrac{13.0945}{2.5819}\right)\right] = 4.5138$

$$Y' = 4.5138X - 4.5138(5.3333) + 78.3333$$

$$Y' = 4.5138X + 54.26$$

X	Y'
0	54.26
5	76.83
10	99.40

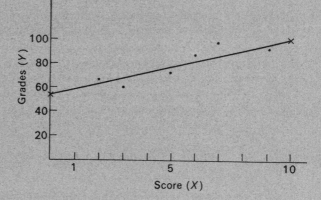

The scattergram indicates that the relationship is fairly linear.

Conclusion   There is a significant ($p < .02$), high, positive correlation ($r = +.89$) between "interest in science" scores and grades in science courses. An increase in ST scores of 1 point is associated with an average grade increase of 4.51 points.

SUMMARY

1. The *Pearson correlation coefficient* is a measure of correlation between two variables based on the relative standing of the subjects within the two groups in terms of Z scores.

2. The *best-fit*, or *regression, lines* represent a scattergram of correlational data in the same sense that the mean represents a set of scores. Regression lines are fit by the method of least squares, so that the sum of squared deviations from the line is minimized. The regression line represents the best guess, or prediction, of a subject's score on one variable on the basis of his score on the other variable. Each scattergram has two best-fit lines, each minimizing the errors in predicting one of the variables.

3. *Errors of estimate*, or prediction, result when the points in the scattergram do not fall on the regression line. The measure of these errors is the *variance* of the points about the regression line. The variance of the errors is a function of the variance of the original scores and an inverse function of the size of the correlation.

4. The existence of a correlation between A and B does *not* mean that A causes or produces B. Evidence in addition to that used in detecting the correlation is needed to show a causal relationship. Manipulating the variables in an experiment is a standard way of investigating possible causal factors.

5. When a subject has been measured for one variable, his score on another variable can be estimated, or *predicted*, if the correlation between the two variables is known. It is assumed that the subject has been drawn from the same population that the sample used to determine the correlation was drawn from. The accuracy of the prediction is given by the error of estimate.

†† A *partial-correlation coefficient* ($r_{12 \cdot 3}$) is a measure of the correlation between factors 1 and 2 when the influence of a third factor (which may affect both 1 and 2) is eliminated or held constant. The partial correlation is a function of the intercorrelations of the three factors.

†† A *multiple correlation* ($R_{1 \cdot 23}$) is the correlation of three or more factors. Predictions can be based on multiple correlations, and the information in additional measures tends to increase the correlation and the accuracy of prediction.

chapter nine

t DISTRIBUTION

In Chapter 7 we saw that when the SD for a sample is calculated from the definitional formula, it tends to underestimate the SD of the population from which the sample was drawn. This bias is minimized if df (i.e., $N - 1$) is used instead of N as a measure of the sample size in calculating s. The Z score (i.e., $(X - \overline{X})/s$) can then be used to answer questions relating to areas under the normal distribution for probabilities of certain events. Even with this correction, however, there is still one important problem. When the sample size is small (theoretically anything less than infinite, but in practice when $N < 100$), the sampling distributions for the Z scores are not quite normal in shape. As the sample size gets smaller, the deviation from normality increases. Basically, the wings of the distribution curve are higher than those for the ND, so that extreme values are somewhat more likely to occur by chance. The characteristics of sampling distributions for different values of df have been calculated, and Table A gives the standard scores needed for selected probability values. When the probability values are taken from the corrected distributions, the standard score as the deviation in SD (s) units is referred to as a "*t* score"* instead of a "Z score." Therefore, although the sampling distribution—a *t* **distribution**—is still considered in terms of standard deviation units, the measure is a *t* score rather than a Z score.

---

* Do not confuse "*t* score" with the T statistic used in the Wilcoxon matched-pairs, signed-ranks test or the T statistic (same name but different measure) used in the White test.

t TEST FOR SINGLE SAMPLES AND SAMPLING DISTRIBUTION OF MEANS

Let us consider a single-sample test in which we wish to determine whether high school students who take an advanced math course differ in ability from the general school population (which has a known mean IQ of 110). We test the math class and find:

$$N = 25 \qquad \bar{X}_{\text{samp.}} = 120$$
$$\bar{X}_{\text{pop.}} = 110 \qquad s = 10$$

How likely is it that a class of 25 students with a mean IQ of 120 (with $s = 10$) is a random sample from a population with a mean IQ of 110? In order to work with a t distribution (as with ND), we need to know the mean and standard deviation of that distribution. We *assume* for the moment that the sample is a random one and the difference between the sample and population means is due to chance. The value of s based on the sample data is then an appropriate measure of the SD of the population. The sampling distribution in which we are interested, however, is not a distribution of scores, but a **distribution of means**. What is the probability that a sample of 25 students will have a mean as high as 120 when the expected mean is 110? To answer this question, we first have to know the characteristics of the sampling distribution for groups of 25 subjects drawn from a population with $\bar{X} = 110$ and $\sigma = 10$. If we generate this distribution by drawing a series of random samples of $N \doteq 25$ from such a population, the means, generally, will be approximately equal to the mean of the population, some higher and some lower. Only occasionally will the obtained means be considerably higher or lower than the population value. The sampling distribution will be essentially normally distributed with the mean ($\bar{X}_{\text{means}} = \bar{X}_{\bar{X}}$) equal to the mean of the population ($\bar{X}_{\text{pop.}}$), which is 110. What would we expect the SD of this sampling distribution of means to be?

Consider two extreme cases: In the first case each sample consists of the score of only one subject. In this case the mean of each sample would be equal to that score (a rather trivial situation), so that the SD of the distribution of means would be the same as that of the s of the individual scores. In the second case each sample includes every member of the population. In this case the sample means would all be identical to the population mean, so that there would not be any variability; the SD of the sampling distribution would be nil. Between these extremes, as the sample size becomes larger, the means of the samples tend to coincide more closely with the mean of the population (since with larger samples there is more opportunity for random, or chance, factors to cancel out). The means of a small sample (e.g., $N = 4$) might, by chance, vary considerably from the mean of the population if the sample happens to include all high-scoring or all low-scoring subjects. With a larger sample (e.g., $N = 25$), it is very unlikely that nearly all the subjects will be at one extreme or the other. So less variability is expected among the means of larger samples than among the means of smaller samples. There is another factor to be considered, and that is the SD of the population itself. If all the scores in the population have a very small range (e.g., between 90 and 110), the sample means will be limited to this range regardless of the sample size. If the SD of the population is very large (e.g., the scores range from 0 to 200), there is much more room for the means to vary by chance for any given sample size. The SD of the sampling distribution of means ($s_{\bar{X}}$) is, therefore, a function of both the population σ (which can be estimated from the sample value of s) and the sample size (N). The relationship turns out, happily, to be a simple one:

$$s_{\bar{X}} = \sqrt{\frac{s^2}{N}} \tag{9.1}$$

where $s^2 = \dfrac{\sum (X - \bar{X})^2}{N - 1} \to \sigma^2_{\text{pop.}}$.

Now we have all the information needed to calculate the value of t for the math class compared to the rest of the school. The data are graphed in Figure 9.1, and

$$s_{\bar{X}} = \sqrt{\frac{10^2}{25}} = \sqrt{\frac{100}{25}} = \sqrt{4} = 2 \tag{9.2}$$

$$t = \frac{\bar{X} - \bar{X}_{\text{pop.}}}{s_{\bar{X}}} = \frac{120 - 110}{2} = \frac{10}{2} = 5 \tag{9.3}$$

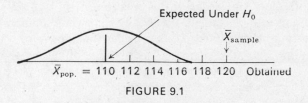

FIGURE 9.1

where $\quad s_{\bar{X}} =$ SD of the distribution of means
$\quad\quad \bar{X}_{\text{pop.}} =$ mean of the population
$\quad\quad\quad \bar{X} =$ mean of the sample

We now need the probability value associated with this statistic (t). In other words, what is the likelihood of getting a t value as large or larger than 5

TABLE 9.1

df	Level of Significance					
	.20	.10	.05	.02	.01	.001
1	3.078	6.314	12.706	31.821	63.657	636.619
2	1.886	2.920	4.303	6.965	9.925	31.598
3	1.638	2.353	3.182	4.541	5.841	12.941
4	1.533	2.132	2.776	3.747	4.604	8.610
5	1.476	2.015	2.571	3.365	4.032	6.859
21	1.323	1.721	2.080	2.518	2.831	3.819
22	1.321	1.717	2.074	2.508	2.819	3.792
23	1.319	1.714	2.069	2.500	2.807	3.767
24	1.318	1.711	2.064	2.492	2.797	3.745
25	1.316	1.708	2.060	2.485	2.787	3.725

df of the t test

minimum t value for
$p = .001$ when $df = 24$

when, by chance, we would expect a value of 0? Table A gives the t values appropriate to selected probability levels. The section of Table A given in Table 9.1 shows that $p < .001$ for $t = 5$ with $df = 24$. The df used for the table is the same as the df in the calculation of s and equals $(N - 1)$. From Table B, $r_m > .61$.

The sample data and the p and r_m values indicate that the math students as a group have an appreciably ($r_m > .61$) and significantly ($p < .001$) higher IQ than the general school population. Whether the students in the math course are brighter because such students are attracted to the course, because they are encouraged (i.e., forced) to take the course, or because the course actually serves to raise the student's IQ are questions that can only be answered with more knowledge of the situation.

Some features of the t statistic are already somewhat familiar, but they are worth discussing briefly. First let us consider the numerator. The value of $\overline{X}_{pop.}$ is the mean of the sampling distribution of means that would be expected if H_0 were true and, therefore, the value that would be expected by chance. Since our example is a single-sample test (one that tests a sample against some prior information or theoretical expected value), the expected value must be obtained from outside the data. This situation is equivalent to that of the χ^2 single-sample test. The sample $\overline{X}$ is the observed value, so that the numerator of the t ratio is a measure of the disparity between what was observed and what would be expected by chance. This is the same type of measure that we used with the χ^2 statistic. In fact, when $df = 1$, χ^2 is algebraically equivalent to t^2. The denominator of the t ratio is a measure of the variability expected by chance, so that the entire ratio is a measure of the difference between the observed and expected values relative to a measure of variability, or the spread of scores within the sample. This ratio shows the general nature of statistical tests with ordinal and interval data. With the White and Wilcoxon tests the simplified calculations served to obscure the principle behind the tests.

The single-sample t test is not used frequently, since it is usually difficult to obtain the necessary information about the population. However, the t test can be extended to permit the comparison of two groups or conditions, and the single-sample t test provides a foundation for an understanding of such a comparison.

t TEST FOR INDEPENDENT SAMPLES AND SAMPLING DISTRIBUTIONS OF DIFFERENCES BETWEEN MEANS

As an example of the testing of differences between the scores of two independent samples, suppose you want to test the effectiveness of two different ways of inducing cooperation in children. A group of 10 four-year-old boys is randomly divided into two groups of 5 boys each. Each boy plays individually in a room in which 10 toys are scattered about the floor. At the end of a 15-minute play session, you ask each child to help clean up the room. All the children get the same request stated in terms of "being a good boy now," "helping me because I'm busy," and other similar appeals to the noble side of a four-year-old boy. As you leave each subject in the experimental group, however, you also mention casually that, if all the toys are on the shelves before you return, you will have a very special present—a toy— for him.* Your measure is the number of toys put away after a five-minute period. The value for $\overline{X}$ and s for each group is shown in Table 9.2.

TABLE 9.2

Number of Toys Put Away by Control Group	Number of Toys Put Away by Experimental Group
0	5
2	7
3	9
4	9
6	10
$\overline{X} = 3, s = 2.2361$	$\overline{X} = 8, s = 2.0$

The null hypothesis for the example is that the two samples can be considered as being drawn from the same population. In other words, if the

---

* A different experimental condition, the induction of stark terror by dire threats, was considered but dismissed as being impractical in an *experimental* situation.

experimental manipulation (offering a toy) does not affect the children's behavior, both sets of scores should be essentially the same. Thus we must compare the means of the two samples and question whether or not it is reasonable that the obtained difference between the means (DM) occurred by chance. (Obtained $DM = \bar{X}_1 - \bar{X}_2 = 8 - 3 = 5$.) To answer this question, we have to know the characteristics of the **sampling distribution of differences between means** that would be expected if H_0 were indeed true. To obtain such a distribution, we could take pairs of samples ($N = 5$ each) from a given population, calculate the mean for each sample, and then determine the difference between the two means. Repeating this process would generate a distribution of the chance values of DM. What would be the characteristics of this distribution? If the two samples are randomly drawn, then there is no reason to expect that the first sample would be consistently larger or smaller than the second sample, so that, on the average, the two samples would be equivalent. In other words, the mean, or average, difference between the means ($\bar{X}_{DM}$) would be equal to 0—the value expected if H_0 is true. With the single-sample t test the expected value had to be obtained from outside the test situation. In the two-sample situation the expected value does not depend upon knowledge of the mean of the population from which the samples were drawn; we do not need any information in addition to the data from the samples. We have, in fact, already seen that with tests comparing two or more samples (e.g., χ^2, White, and Kruskal-Wallis tests) outside information is not required to determine the values expected by chance.

The t test involves the assumption that both samples are drawn from populations that are normally distributed and have the same variability (i.e., $\sigma^2_{\text{pop } 1} = \sigma^2_{\text{pop. 2}}$). In terms of H_0, we assume that both samples are drawn from the same normally distributed population. In practice, it is difficult or impossible to determine if the samples are actually drawn from a normally distributed population, but generally we are satisfied if the data are not highly skewed and the $\bar{X}$ and s of each sample are suitable measures. The values of s for each sample do not have to be equal, but as the difference between sample s values increases, the efficiency, or sensitivity, of the test decreases.

We now need the value of the SD for the sampling distribution of differences between means (s_{DM}). Would we expect large or only small differences between pairs of means? The answer clearly depends upon the likelihood of the means varying from one another by chance. The variability expected for the distribution of sample means is given by $s_{\bar{X}}$ (formula 9.1). On the assumption that the samples are drawn from populations having equivalent variance, combining the values of $s_{\bar{X}}$ for the two samples will give an adequate measure for the SD of the sampling distribution. This measure is usually referred to as the standard deviation or standard error of the difference between means (s_{DM}). The value of s_{DM} can be obtained by using the following formula (for equal sample sizes):

$$s_{DM} = \sqrt{s_{\bar{X}1}^2 + s_{\bar{X}2}^2} \qquad (9.4)$$

when $s_{\bar{X}1}^2 = \dfrac{s_1^2}{N}$

since $s_{\bar{X}1} = \dfrac{s_1}{\sqrt{N}}$

We can now calculate the t value using the following formula:

$$t = \frac{DM - \bar{X}_{DM}}{s_{DM}} = \frac{DM - 0}{s_{DM}} = \frac{DM}{s_{DM}} \qquad (9.5)$$

For the example, we have the data given in Table 9.3 and the following calculations:

$$s_{DM} = \sqrt{1 + .8} = \sqrt{1.8} = 1.3416 \qquad (9.6)$$

$$t = \frac{DM - \bar{X}_{DM}}{s_{DM}} = \frac{(8 - 3) - 0}{1.3416} = \frac{5}{1.3416} = 3.73 \qquad (9.7)$$

The df for the test is the sum of the degrees of freedom for the two samples or $(5 - 1) + (5 - 1) = 8$. For $t = 3.73$, $df = 8$, $p < .01$ (from Table A), and $r_m > .76$ (from Table B). Therefore, we have a fairly large ($r_m > .76$) and significant ($p < .01$) difference in performance between the experimental and control groups, with the children that were offered a reward tending to put away more toys ($\bar{X} = 8$ out of 10) than those in the control group ($\bar{X} = 3$ out of 10).

TABLE 9.3

Control Group	Experimental Group
$\sum X = 15$	$\sum X = 40$
$\sum X^2 = 65$	$\sum X^2 = 336$
$\bar{X} = \dfrac{15}{5} = 3$	$\bar{X} = \dfrac{40}{5} = 8$
$s^2 = 5.0$	$s^2 = 4.0$
$s_{\bar{X}}^2 = 1$	$s_{\bar{X}}^2 = .8$

These same data could have been analyzed in the form of ranks with a White test, which would have led to a T value of 16 for which $p < .05$ and $r_m > .63$. The t test, then, results in a lower probability value and a larger

r_m value than the White test. In general, when the t test is suitable for the data, we are more apt to detect differences that are really there (i.e., to avoid type II errors) with the t test than the White test. In other words, when the assumptions of the t test are met, the t test is more sensitive or more powerful than the White test.

The value for the s_{DM} is obtained by pooling values for $s_{\bar{X}}$ for the two samples. When the two samples do not have an equal number of subjects, the formula has to be adjusted to give greater weight to the sample with more subjects. As with the standard deviation, formulas algebraically equivalent to the definitional formula can be stated in terms of raw scores. The raw-score formulas for s_{DM} eliminate rounding errors as well as being highly suited for use with a calculator or computer.

When the sample sizes are not equal, the following formula is appropriate:

$$s_{DM} = \sqrt{\frac{(\sum X_1^2 + \sum X_2^2) - (n_1 \bar{X}_1^2 + n_2 \bar{X}_2^2)}{n_1 + n_2 - 2}\left(\frac{1}{n_1} + \frac{1}{n_2}\right)} \qquad (9.8)$$

When the two sample sizes are equal ($n_1 = n_2$), formula (9.8) reduces to:

$$s_{DM} = \sqrt{\frac{(\sum X_1^2 + \sum X_2^2) - n(\bar{X}_1^2 + \bar{X}_2^2)}{n(n-1)}} \qquad (9.9)$$

TABLE 9.4

Control Group	Experimental Group
$\sum X = 15$	$\sum X = 40$
$\bar{X} = 3$	$\bar{X} = 8$
$\sum X^2 = 65$	$\sum X^2 = 336$

Using the data from the example (which is repeated in Table 9.4) with formulas (9.8) and (9.9), we get the same value for s_{DM} that was obtained in equation (9.6):

$$s_{DM} = \sqrt{\frac{(65 + 336) - 5(3^2 + 8^2)}{5(4)}} \qquad (9.10)$$
$$= 1.3416$$

or

$$s_{DM} = \sqrt{\frac{(65 + 336) - [5(3^2) + 5(8^2)]}{5 + 5 - 2}\left(\frac{1}{5} + \frac{1}{5}\right)} \qquad (9.11)$$

$$= \sqrt{\frac{36}{8}(.2 + .2)} = \sqrt{4.5(.4)} \tag{9.11}$$

$$= 1.3416$$

Assume that we now modify the experiment of our example by placing 25 toys on the floor. Instead of telling the experimental group to put all the toys away, we mention only that they will each get a present if they do a "very good job." We obtain the data of Table 9.5 and can calculate both the t test:

$$t = 3.51 \qquad DM = 4 \qquad s_{DM} = 1.14$$
$$p < .01 \qquad r_m > .76$$

and the White test:

$$T = 16$$
$$p < .05 \qquad r_m > .63$$

Once again the t test for these data is more efficient than the White test.

TABLE 9.5

Control Group		Experimental Group	
Number of Toys Put Away	Rank	Number of Toys Put Away	Rank
2	1.5	4	5
2	1.5	6	7
3	3.5	7	8
3	3.5	8	9
5	6	10	10
$\bar{X} = 3$	$\sum R = 16$	$\bar{X} = 7$	$\sum R = 39$

But what happens to the situation if the best subject in the experimental group (with a score of 10) becomes quite manic (wild) at the prospect of getting a present and cleans up *all* 25 toys? If all the other scores remain the same, this change from 10 to 25 in that single score will certainly raise the mean for the experimental group and increase the value of *DM*. In other words, the difference between the groups will be larger than before. Using the data of Table 9.6, we can see the effect of this change on both the t test:

$$t = 1.82 \qquad DM = 7 \qquad s_{DM} = 3.85$$
$$.10 > p > .05 \qquad r_m < .63$$

and the White test:

$$T = 16$$
$$p < .05 \qquad r_m > .63$$

TABLE 9.6

Control Group		Experimental Group	
Number of Toys Put Away	Rank	Number of Toys Put Away	Rank
2	1.5	4	5
2	1.5	6	7
3	3.5	7	8
3	3.5	8	9
5	6	25	10
$\bar{X} = 3$	$\sum R = 16$	$\bar{X} = 10$	$\sum R = 39$

What happened? Even though the difference between the two groups is now larger, the t test no longer indicates a significant difference, and the White test (with $p < .05$) is unaffected. The increased value of DM is more than offset by the very large increase in s_{DM}. This is the result of the great increase in s for the experimental group. The extreme score (the 25) increases the mean slightly but causes a very large increase in s, and consequently in s_{DM}. The data for the experimental group are now highly skewed and, as discussed before, the mean and standard deviation are not appropriate measures for a skewed distribution. The White test is unaffected by this change, since the highest score has the highest rank whether the value is 10 or 25. The insensitivity of the ranking test to the absolute values of the scores also tends to make the ranking test less sensitive to the effects of skewed distributions. For the data in Table 9.6 the more appropriate test is the White test, since the t test is weakened when the data are skewed and the sample size is small. As sample sizes increase, the t test is less affected by violations of the assumptions that the samples are from normal populations and have equal variance.

In Chapter 7 we saw that the mean and the median can be compared to determine whether or not a distribution is skewed, and if it is, the median might be a better measure than the mean. In similar crude fashion we can compare the outcomes of the t test with those of the comparable ranking test, and if the ranking test shows a difference that the t test does not and if the data appear to be inappropriate for a *t* test, we would base the interpretation of an experiment on the ranking test. This problem will be discussed in a broader context in Chapter 13.

t TEST FOR RELATED SAMPLES

When using related samples (in either a repeated-measures or matched-pairs design), we are not particularly concerned about the overall difference between the scores of the two conditions but rather with the consistent differences between the related pairs of scores. Previously we dealt with this situation for ordinal data with the Wilcoxon signed-ranks test. The *t* test can also be used for related-samples tests, in exactly the same way it is used for independent samples except that the value for s_{DM} has to be changed to account for the use of related samples. This value is given by:

$$s_{DM} = \sqrt{(s_{\bar{X}_1})^2 + (s_{\bar{X}_2})^2 - 2r_{1 \cdot 2}(s_{\bar{X}_1})(s_{\bar{X}_2})} \qquad (9.12)$$

The term $r_{1 \cdot 2}$ is the correlation coefficient between the two scores for each matched pair.

If the two scores for each pair are correlated—that is, if pairs of subjects are matched on the basis of a relevant variable, or if, in situations when the same subject is tested twice, sequential effects do not produce inconsistent results—then the value of s_{DM} will be smaller than when independent groups are used. Therefore, with a given value for DM, the value for the *t* ratio (DM/s_{DM}) tends to become larger as the correlation $(r_{1 \cdot 2})$ increases. As an extreme example consider what happens when $s_{\bar{X}_1} = s_{\bar{X}_2}$ and $r = 1$:

$$s_{DM} = \sqrt{(s_{\bar{X}_1})^2 + (s_{\bar{X}_2})^2 - 2(1)(s_{\bar{X}_1})(s_{\bar{X}_2})} = \sqrt{2(s_{\bar{X}})^2 - 2(s_{\bar{X}})^2} = \sqrt{0} = 0 \qquad (9.13)$$

If the correlation is perfect, if the relative standing of each individual's score is identical on both tests, then any difference between the overall level of scores of two samples would be significant.

Consider, for example, the sets of data in Table 9.7; in case 1 the data are from an independent-samples design, and in cases 2 and 3 they are from related samples. With the independent samples the difference between the conditions is relatively small compared to the spread of scores within each group, and with this sample size the difference between groups is not significant. When we obtain the same scores with matched pairs of subjects, however, there is clearly a tendency for the scores to increase within every pair. The difference between the two samples is certainly more visible in the related-samples situation, and in fact, a Wilcoxon test would show the difference nicely. Note that in case 2 the order of the subjects is the same

TABLE 9.7

Case 1, *Independent Samples*			Case 2, *Related Samples*			Case 3, *Related Samples*		
A	B		Subjects	A	B	Subjects	A	B
2			a	2		a	2	
4			b	4		b	4	
	5				5			5
8	8		c	8	8	c	8	8
10	10		d	10	10	d	10	10
	14				14			14
15			e	15		e	15	
	16				16			16
17			f	17		f	17	
18			g	18		g	18	
	20				20			20
	22				22			22
$t = .89$			$t = 6.87$			$t = 1.01$		
$p > .20$			$p < .001$			$p > .20$		
$r_m < .20$			$r_m > .80$			$r_m < .20$		

under both the A and B conditions.* In case 3, the overall differences between A and B are the same as in 1 and 2, but now the related-samples test shows no consistent pattern of difference between the matched subjects. The order of the subjects under conditions A and B differs greatly, and the value of the correlation is essentially 0. Thus, the increased efficiency of the related-samples design over the independent-samples design depends on the adequate matching of subjects in the two groups or the reliability (consistency) of subjects tested twice.

The calculation of the related-samples t test can be carried out with formula

* This ordering would give a perfect correlation ($r = +1$) for the Spearman rank-order correlation, but the relationship is not necessarily perfect if we use the interval properties of the scores for a Pearson correlation.

(9.12) for s_{DM}, which includes a correlation coefficient, but this is a cumbersome and rarely used procedure. Algebraically equivalent ways of calculating the *t* value require considerably less effort.

TABLE 9.8

Subject	Score under Quiet Conditions	Score under Noisy Conditions	Difference (D)	D^2
a	25	15	−10	100
b	24	10	−14	196
c	22	12	−10	100
d	19	13	−6	36
e	19	6	−13	169
f	16	19	+3	9
g	15	7	−8	64
h	13	14	+1	1
i	12	4	−8	64
j	10	5	−5	25
			$\sum D = -70$	$\sum D^2 = 764$

Look again at the data analyzed by the Wilcoxon signed-ranks test in Chapter 4, which is repeated in Table 9.8. In the ranking test we deal only with the difference between the scores within each pair. The original scores for each pair are ignored, and attention paid to only the set of difference scores. If the null hypothesis is true and performance under both conditions is essentially the same, what do we expect for each pair of scores? The scores within each pair or, when each subject is tested twice, for each subject would not necessarily be the same under both conditions, since there are always various chance factors. Nevertheless, there is no reason for either the first or second set of scores to be consistently higher than the other. In fact, in the Wilcoxon test (+) and (−) differences are equally likely within the sample of difference scores. Similarly, large and small differences are equally likely in both directions. If the two conditions are essentially the same, then, the differences in each direction should, on the average, cancel out. With interval data we would similarly expect a mean difference ($\bar{X}_D$) of 0. In Table 9.8 the mean difference is 7.0. Could $\bar{X}_D = 7.0$ reasonably occur by chance under two truly equal conditions? Ignore for the moment that these are difference scores, and consider the data as just one set of scores, a situation which is exactly the one appropriate for the *t* test for a single sample. In this case, however (in contrast to the usual single-sample test), we do *not* need any outside information to find the mean value of the distribution of chance mean differences ($\bar{X}_{\text{mean difference}} = \bar{X}_{\bar{D}}$), since, if H_0 is correct, $\bar{X}_{\bar{D}} = 0$.

We can now use the procedure given in formulas (9.1) and (9.3) to obtain the value for the t ratio. The s derived from the set of scores (difference scores in this case) is still the best estimate of $\sigma_{pop.}$, and $s_{\bar{D}}$ provides an adequate estimate of the sampling distribution for mean differences ($\bar{X}_D$). Thus, we can easily carry out the calculation, calling each score "D" instead of "X" to denote difference scores:

$$(s_D)^2 = \frac{N \sum D^2 - (\sum D)^2}{N(N-1)} = \frac{10(764) - (70)^2}{10(9)} = 30.4444 \qquad (9.14)$$

$$s_{\bar{D}} = \sqrt{\frac{(s_D)^2}{N}} = \sqrt{\frac{30.4444}{10}} = \sqrt{3.04444} = 1.7448 \qquad (9.15)$$

$$t = \frac{\bar{D} - \bar{X}_D}{s_{\bar{D}}} = \frac{7 - 0}{1.7448} = \frac{7}{1.7448} = 4.01 \qquad (9.16)$$

The value for df in this case is the number of difference scores minus one ($df = N - 1$). With the t test for independent samples $df = (n_1 - 1) + (n_2 - 1)$, so for the present data (20 observations) $df = 18$. With the related-samples test $df = 18/2 = 9$. Since larger values of t are needed for significance as df decreases, this reduced df makes it harder to reject the null hypothesis. Where did the missing df go? The reduced df is certainly appropriate for the single-sample test (where 1 df is lost by fixing the value of $\bar{X}_D$). In formula (9.12) for s_{DM} based on the correlation coefficient, degrees of freedom are lost not only for fixing the means of each group but also for fixing the correlation between the two sets of scores. The net df ($N - 1$) is the value used.

With $t = 4.01$, we find that $p < .01$ and $r_m > .85$, which is essentially the same outcome that we had for these data with the Wilcoxon test. The conclusion is that the scores tend to be appreciably ($r_m > .85$) and significantly ($p < .01$) higher under the quiet condition than under the noisy condition.

The same problems connected with the use of related samples apply here as for the Wilcoxon test. The related-samples t test is more efficient and sensitive than the Wilcoxon test, except when the distribution of difference scores is so skewed that the $\bar{X}_D$ and s_D are not suitable measures of the difference scores.

Calculations for the related-samples t test can be simplified with the following formula:

$$t = \sqrt{\frac{N - 1}{\left[\dfrac{N \sum D^2}{(\sum D)^2}\right] - 1}} \qquad (9.17)$$

which for the data in Table 9.8 yields the following value:

$$t = \sqrt{\frac{N-1}{\left[\frac{N \sum D^2}{(\sum D)^2}\right] - 1}} = \sqrt{\frac{10-1}{\left[\frac{10(764)}{(70)^2}\right] - 1}} = \sqrt{\frac{9}{\left[\frac{7,640}{4,900}\right] - 1}} \qquad (9.18)$$

$$= \sqrt{\frac{9}{1.559 - 1}} = 4.01$$

Formula (9.17) involves fewer steps than the procedure in formulas (9.14), (9.15), and (9.16), but the value for the average difference, $\bar{D} = (\sum D)/N$, must still be calculated before the situation can be properly interpreted.

The t test is the most sensitive and commonly used procedure to compare two conditions, and this chapter does not conclude the discussion of this test. The relationship between analysis of variance and the t test will be shown in Chapters 10 and 11. In Chapter 12 the t test will be brought up again with reference to measuring the size of the observed difference (using r_m) and with regard to selecting an appropriate sample size. In Chapter 13, the problem of choosing between the t test and a ranking test will be dealt with again.

ONE- AND TWO-TAILED PROBABILITY VALUES

Assume that t values are calculated for the difference between the means of two random samples from the same population. Given that the null hypothesis is true, the sampling distribution of such t values would be essentially normally distributed (when N is large). With a 5 percent significance level the proportion of chance t values that would lead to rejection of the null hypothesis (type I error) would be no larger than 5 percent. This is equivalent to considering all values of Z beyond ± 1.96 significant. The region of rejection (the area containing the significant outcomes) is shown in the sampling distribution of t in Figure 9.2.

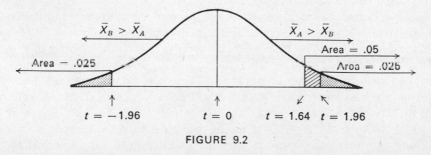

FIGURE 9.2

The shaded area of the normal distribution represents 5 percent of the total area and defines the level of significance. Note that cases leading to rejection of H_0 are found at both extremes (or both tails) of the distribution. In other words, a difference between the means of the two samples in *either* direction (i.e., $\bar{X}_A > \bar{X}_B$ or $\bar{X}_B > \bar{X}_A$) associated with a Z value of ± 1.96 or larger would lead us to reject H_0. Probability values based on the possibility of outcomes in either direction are called **two-tailed values**. The probability tables for most of the statistical measures given in this text are in terms of two-tailed values.

What about situations in which we are concerned with differences in only one direction and not in the other? For example, suppose we will use a new text or a new medicine *only* if it is better than the one we are currently using. If it is not better, we do not care whether it is the same as or worse than the current text or treatment. The statistical problem is to determine the portion of the sampling distribution representing improvements in performance that would occur by chance no more than 5 percent of the time. In other words, when the only differences that would lead to rejection of H_0 are outcomes in one direction, the area of rejection is confined to one side or one tail of the sampling distribution, as shown by the partially shaded area in Figure 9.2. Probability values based on possible outcomes in only one direction are called **one-tailed values**. Note that, although a t value of 1.96 is required to reject H_0 at the 5 percent level with a two-tailed test, a t value of only 1.64 is needed if a one-tailed test is used. Many investigators find it tempting to use a one-tailed probability level to facilitate obtaining "significant" results.

The choice between one- and two-tailed probability values is the subject of considerable controversy in statistical and research literature, and the main points are summarized here:

1. With a given t value, it is easier to reject the null hypothesis with a one-tailed than with a two-tailed probability level. Therefore, the chances of making a type I error (mistakenly rejecting H_0) are greater with a one-tailed test, and the chances of making type II errors are greater with the two-tailed test. Using a one-tailed test with a .05 probability level is exactly equivalent to using a two-tailed test with a .10 probability level. Therefore, the relative danger and cost of type I and type II errors can be determined on the basis of using a lenient level of significance.

2. The magnitude of effect measure (r_m) is important to the interpretation of the data. As we shall see in Chapter 12, r_m is a function of the value of the statistic (e.g., t) and the sample size, but *not* of the probability value directly. In other words, the r_m value for a given t test is the same whether a one- or a two-tailed probability value is used. Therefore, using a one-tailed value may make it easier to reject H_0, but it will not affect the final interpretation of the size of the effect.

3. A major problem with the use of one-tailed values is what to do with results that are in the opposite direction of those expected. Theoretically, once an investigator decides to use one-tailed values, outcomes in the opposite (nonpredicted) direction should be ignored. However, overlooking large and interesting differences is often unwise, and in practice it is certainly very hard to do so. Assume, for example, that we are testing a new text.

Presumably, we are interested only in results showing that the new text is better than the old, but to our surprise, the new text is much worse. Some apparently minor aspects of the texts must be responsible for the large difference obtained. We decide to test to see if the new text is significantly *worse* than the old text, and to be honest (since we did not predict the direction of the outcome), we use two-tailed probability values and reject H_0 at the .05 level. Under these conditions, what is the actual chance of rejecting H_0 at the .05 level? If the results had been in the anticipated direction, there would have been a .05 chance of rejecting H_0; that is, if the t value had fallen within 5 percent of the area confined to the right-hand tail of the distribution, H_0 would have been rejected. But the results were in the opposite direction, and since we are using two-tailed probability values, there is a .025 probability of rejecting H_0; if the t value falls in the 2.5 percent area in the left-hand tail of the distribution, H_0 will be rejected. Adding these areas together results in a $.05 + .025 = .075$ probability of rejecting the null hypothesis at the .05 level—a disturbing outcome. We *cannot* plan to use one-tailed values for results in the predicted direction and then use two-tailed values for results in the other direction—in fact, there is no logically consistent way of testing for results in the nonpredicted direction after deciding to use one-tailed values.

In view of these three considerations, the safest procedure in virtually all situations is to use two-tailed values. Using one-tailed values to make rejection of H_0 "easier" serves to increase type I errors, while the size of the difference, as measured by r_m, is unaffected. A strict application of one-tailed values includes the danger of overlooking important results in the non-predicted direction, and any attempt to test for such outcomes leads to inaccurate probability values. Analysis of data based on two-tailed probability levels can be reported without apology, while it is almost always necessary to "explain away" the use of one-tailed probability levels.

A minor but related problem arises when statistics with markedly skewed sampling distributions are based on data from small samples. The binomial test when $p \neq q$ and the Fisher test for the 2×2 contingency table (cf. Siegel 1956) are measures in which deviations from H_0 in one direction do not have the same likelihood as deviations in the other direction. Here a one-tailed probability value and a more stringent level of significance (e.g., .02 instead of .05) is sometimes used. A simpler alternative is to approximate two-tailed values by doubling the one-tailed probabilities so that changes in both directions are considered, though the region of rejection in the opposite direction is not necessarily specified.

t TEST FOR SINGLE SAMPLES

Formula

$$t = \frac{\bar{X}_o - \bar{X}_e}{s_{\bar{X}}}$$

where $df = N - 1$

$$s_{\bar{X}} = \sqrt{\frac{s^2}{N}}$$

$$s^2 = \frac{N \sum X^2 - (\sum X)^2}{N(N-1)}$$

$\bar{X}_o$ = observed sample mean = $\dfrac{\sum X}{N}$

$\bar{X}_e$ = assumed population mean and expected sample mean

Table A   provides p values based on t and df.

Table C   provides r_m values based on t and df.

Restrictions   The expected mean value ($\bar{X}_e$) is based on prior knowledge of the situation or derived from theoretical considerations. The sample should be reasonably normally distributed with the mean and standard deviation appropriate measures.

Procedure

1. Determine the value of $\bar{X}_e$ from prior knowledge of the assumed population or from a theory.
2. Sum each score to obtain $\sum X$.
3. Square each score and sum the squares to obtain $(\sum X)^2$. Note: The terms $\sum X$ and $(\sum X)^2$ can be obtained simultaneously with a single entry of each score on most modern calculators.
4. Calculate $\bar{X}_o$ by dividing $\sum X$ by N.
5. Calculate s^2 by substituting $\sum X$, $(\sum X)^2$, and N in the formula.
6. Divide s^2 by N and take the square root to obtain $s_{\bar{X}}$.
7. Calculate the t value by subtracting $\bar{X}_e$ from $\bar{X}_o$ and dividing the difference by $s_{\bar{X}}$.

8. Use Table A to obtain p values based on t and $df = N - 1$.
9. Use Table C to obtain r_m values based on t and $df = N - 1$.

Example

Previous studies have shown that the average mouse gains 5 grams between weaning at 21 days and 30 days, when fed a standard diet. A group of 20 just-weaned mice is given food sweetened with saccharin, and the weight gain of each mouse is measured. The weight gain, in grams, of each mouse in the sample is as follows:

4.8	5.4	5.1	5.4
5.1	6.0	5.2	5.0
4.9	5.2	4.5	5.9
5.8	4.7	5.4	4.6
5.3	5.6	4.7	6.2

$$N = 20 \qquad \sum X = 104.8 \qquad \sum X^2 = 553.56$$

$$\bar{X}_e = 5.0$$

$$\bar{X}_o = \sum X/N = 5.24$$

$$s^2 = \frac{N \sum X^2 - (\sum X)^2}{N(N - 1)} = \frac{20(553.56) - (104.8)^2}{20(19)}$$

$$= \frac{88.16}{380} = .2320$$

$$s_{\bar{X}} = \sqrt{\frac{s^2}{N}} = \sqrt{\frac{.2320}{20}} = \sqrt{.01160} = .1077$$

$$t = \frac{\bar{X}_o - \bar{X}_e}{s_{\bar{X}}} = \frac{5.24 - 5.0}{.1077} = \frac{.24}{.1077} = 2.228$$

From Table A, based on $t = 2.228$ and $df = 20 - 1 = 19$, $p < .05$.

From Table C, based on $t = 2.228$ and $df = 19$, $r_m > .45$.

Conclusion

The mean weight gain of the sample fed sweetened food is significantly ($p < .05$) and moderately greater ($r_m > .45$) than the mean weight gain of the reference group.

t TEST FOR INDEPENDENT SAMPLES

Formula

$$t = \frac{\bar{X}_1 - \bar{X}_2}{s_{DM}}$$

where $df = n_1 + n_2 - 2$.
When $n_1 = n_2$ (equal sample sizes):

$$s_{DM} = \sqrt{\frac{(\sum X_1{}^2 + \sum X_2{}^2) - n(\bar{X}_1{}^2 + \bar{X}_2{}^2)}{n(n-1)}}$$

When $n_1 \neq n_2$ (unequal sample sizes):

$$s_{DM} = \sqrt{\frac{(\sum X_1{}^2 + \sum X_2{}^2) - (n_1\bar{X}_1{}^2 + n_2\bar{X}_2{}^2)}{n_1 + n_2 - 2} \left(\frac{1}{n_1} + \frac{1}{n_2}\right)}$$

Table A provides p values based on t and df.

Table C provides r_m values based on t and df.

Restrictions The mean and standard deviation should be appropriate measures for each sample. The variances of the two samples should be approximately equal. The effect of skewness, or unequal variance, decreases as the sample size increases. When one or both samples are highly skewed, the White test may be more suitable.

Procedure

1. For each sample determine $\sum X$, $\sum (X^2)$, and n.
2. Divide $\sum X_1$ by n_1 and $\sum X_2$ by n_2 to obtain $\bar{X}_1$ and $\bar{X}_2$.
3. Calculate s_{DM}. If $n_1 = n_2$, use the simpler formula.
4. Divide $(\bar{X}_1 - \bar{X}_2)$ by s_{DM} to obtain t.
5. Enter t and df in Table A to obtain p.
6. Enter t and df in Table C to obtain r_m.

Example Two groups of students are given math problems of equivalent difficulty. For one group the problems require addition, and for the other group they require subtraction. The number of problems solved in a two-minute test session is shown below:

Addition Problems	$X_1{}^2$	Subtraction Problems	$X_2{}^2$
15	225	9	81
12	144	6	36
10	100	5	25
8	64	5	25
8	64	4	16
$\sum X_1 = 53$	$\sum X_1{}^2 = 597$	$\sum X_2 = 29$	$\sum X_2{}^2 = 183$

$$\bar{X}_1 = 10.6 \qquad\qquad \bar{X}_2 = 5.8$$

For $n_1 = n_2$,

$$s_{DM} = \sqrt{\frac{(597 + 183) - 5[(10.6)^2 + (5.8)^2]}{5(5 - 1)}}$$

$$s_{DM} = \sqrt{\frac{780 - 5(146)}{20}} = \sqrt{\frac{50}{20}} = 1.5811$$

Or for $n_1 \neq n_2$,

$$s_{DM} = \sqrt{\frac{(597 + 183) - [5(10.6)^2 + 5(5.8)^2]}{5 + 5 - 2}\left(\frac{1}{5} + \frac{1}{5}\right)}$$

$$= \sqrt{\frac{780 - [5(112.36) + 5(33.64)]}{8}\left(\frac{2}{5}\right)}$$

$$= \sqrt{\frac{50}{8}\left(\frac{2}{5}\right)} = \sqrt{\frac{100}{40}} = 1.5811$$

$$t = \frac{DM}{s_{DM}} = \frac{10.6 - 5.8}{1.5811} = \frac{4.8}{1.5811} = 3.036$$

From Table A, based on $t = 3.036$ and $df = 8$, $p < .01$.

From Table C, based on $t = 3.036$ and $df = 8$, $r_m > .70$.

Conclusion   The mean number of problems solved by the group doing addition is appreciably ($r_m > .70$) and significantly ($p < .01$) greater than the number solved by the group doing subtraction.

_t_ TEST FOR RELATED SAMPLES

Formula

(1) $t = \dfrac{\bar{D}}{s_{\bar{D}}}$

where $df = N - 1$

$$s_{\bar{D}} = \sqrt{\frac{(s_D)^2}{N}}$$

$$(s_D)^2 = \frac{N \sum D^2 - (\sum D)^2}{N(N-1)}$$

(2) or $t = \sqrt{\dfrac{N-1}{\left[\dfrac{N \sum D^2}{(\sum D)^2}\right] - 1}}$

Table A   provides _p_ values based on _t_ and _df_.

Table C   provides r_m values based on _t_ and _df_.

Restrictions The distribution of difference scores should be reasonably normal, so that the mean difference ($\bar{D}$) is an appropriate measure. If the difference scores are highly skewed, the Wilcoxon matched-pairs signed-ranks test may be more suitable.

Procedure

1. Subtract the first score for each subject (or matched pair) from the second score to obtain the difference score (D). Retain the sign of the D score.
2. Add the D scores across all subjects (or matched pairs) to obtain $\sum D$.
3. Square each D score and add to obtain $\sum D^2$.
   Note: The terms $\sum D$ and $\sum D^2$ can be generated simultaneously with a single entry of each D score on most modern calculators.
4. Divide $\sum D$ by N (number of subjects or matched pairs) to obtain the mean difference($\bar{D}$).
5. Substitute N, $\sum D$, and $\sum D^2$ in the formula to calculate the variance of the differences $(s_D)^2$.
6. Divide $(s_D)^2$ by N and find the square root to obtain the standard error of the mean differences $s_{\bar{D}}$.
7. Divide $\bar{D}$ by $s_{\bar{D}}$ to determine the t score.
8. Use Table A to determine the p value based on t and $df = N - 1$.
9. Use Table C to determine the r_m value based on t and _df_.

or 4a. Calculate the t value directly using formula (2). Calculate the value of $\bar{D}$ separately. This method may be quicker than steps 4–7.

Example

Student performance on a mental agility test is measured at the beginning and end of a course in basic statistics.

Subject	At Beginning of Course	At End of Course	D	D^2
a	80	95	+15	225
b	70	60	−10	100
c	60	65	+5	25
d	50	75	+25	625
e	45	60	+15	225
f	40	55	+15	225
g	25	45	+20	400
$N = 7$			$\sum D = 85$	$\sum D^2 = 1{,}825$

$$\bar{D} = \frac{\sum D}{N} = \frac{85}{7} = 12.14$$

$$(s_D)^2 = \frac{7(1{,}825) - (85)^2}{7(6)} = \frac{5{,}550}{42} = 132.1428$$

$$s_{\bar{D}} = \sqrt{\frac{132.1428}{7}} = \sqrt{18.8775} = 4.3448$$

$$t = \frac{\bar{D}}{s_{\bar{D}}} = \frac{12.14}{4.3448} = 2.79$$

$$\text{or } t = \sqrt{\frac{N-1}{\left[\dfrac{N\sum(D^2)}{(\sum D)^2}\right] - 1}} = \sqrt{\frac{6}{\left[\dfrac{7(1{,}825)}{(85)^2}\right] - 1}}$$

$$= \sqrt{\frac{6}{(1.7682) - 1}} = \sqrt{\frac{6}{.7682}} = 2.79$$

From Table A, $p < .05$ for $t = 2.79$ with $df = 6$.

From Table C, $r_m > .75$ for $t = 2.79$ with $df = 6$.

Conclusion

The mental agility scores were significantly ($p < .05$) and appreciably ($r_m > .75$) higher at the end of the statistics course than at the start of the course.

SUMMARY

1. The sampling distribution of Z scores [i.e., $(X - \bar{X})/s$] is not quite normal in form when sample sizes are small. Corrected probability values appropriate to the sample size (in terms of df) can be obtained from the sampling distribution. When the correction has been made, the Z score is called a *t score* and the sampling distribution is called a *t distribution*.

2. If samples are drawn from a population and the mean calculated for each sample, the obtained means will produce a *sampling distribution of means*. The mean of the distribution is equal to the mean of the population (i.e., $\bar{X}_{\bar{X}} = \bar{X}_{\text{pop.}}$); the SD of the distribution is a direct function of the population SD, an inverse function of sample size.

3. The *t test for single samples* is a test to determine if the obtained $\bar{X}$ for a sample is likely for a sample obtained from a population having a known or theoretical mean value. The obtained $\bar{X}$ score is compared to the sampling distribution of means that would be expected if H_0 were correct.

4. A *sampling distribution of differences between means* can be obtained by repeatedly taking *pairs* of samples, calculating the means for each sample and the difference between means for each pair. These differences will generate the distribution. The mean of the distribution is equal to 0 (i.e., $\bar{X}_{DM} = 0$), and the SD (s_{DM}) is a direct function of the SD of the samples and an inverse function of the size of the samples (N).

5. The *t* test for independent samples is analogous to the White test, but, when the data are suited to the use of means and standard deviations, the *t* test is more sensitive. The obtained DM is compared to the sampling distribution for differences between means.

6. The *t* test for related samples is similar to the *t* test for independent samples, but the value of s_{DM} is reduced, and the size of *t* increased to take into account the use of repeated measures or matched pairs. The Wilcoxon matched-pairs signed-ranks test is equivalent but less sensitive. Another method of calculating the test, which is algebraically simple, is based on the difference between scores within pairs. The obtained *t* value is compared to the appropriate *t* distribution.

7. Probability values that allow for deviation from the expected value in either direction (positive or negative) are known as *two-tailed probability values*. When outcomes in one direction only are to be considered, *one-tailed probability values* can be employed to decrease the probability for a given value of the statistic and thus make it easier to reject H_0. Two-tailed values are generally used, since serious problems exist when one-tailed values are used.

chapter ten

We previously analyzed the data from more than two independent samples by extending the ranking test for two samples to form the Kruskal-Wallis test. When applied to two groups, the Kruskal-Wallis test is essentially identical to the White test, and the White test may be considered a special, simplified version of the more general procedure for analyzing data from two or more samples. In an analogous fashion the t test for two independent samples can be extended to handle k samples. Although the initial analysis will not look like a version of the t test, the relationship will become clear by the end of the chapter.

THE F RATIO

In Chapter 7 samples were described in terms of the mean and standard deviation. The means of two samples were compared to the sampling distribution of the t ratio in order to determine whether or not the two samples could reasonably have been drawn from the same population. In a similar way the variance (s^2) can be used as a basis for deciding if two samples differ only by chance. The appropriate statistic is the ratio of the variances (i.e., s_1^2/s_2^2), which for the multi-group situation provides a measure analogous to the t ratio, and this ratio of variances is called the F ratio. The main characteristics of the sampling distribution of this ratio can readily be estimated. If we were to take random samples from the same population, calculate the variances of these samples, and determine the F ratio for each pair of samples, what distribution would be formed? Since the s^2 of each sample is an estimate of the variance of the population from which the

155

sample was drawn, the values of s^2 for each sample will tend to be approximately the same. It would not be reasonable for the first sample drawn to have consistently higher or lower values of s^2 than the second sample. Therefore, the mean value expected by chance for the F ratio is 1.0, and values increasingly larger or smaller are increasingly unlikely. To be precise, the expected value is slightly greater than 1.0 when the df of the denominator is small and approaches 1.0 as the df increases. The sampling distribution of F is also a function of the df of the two samples, since as the sample sizes increase, the sample variances approximate the population variance and each other better. Tables giving the probability values associated with the F ratio (Table E) are in terms of the df for both the numerator (top) and the denominator (bottom). In practice if we were to compare two samples using the F ratio and find that the likelihood of getting an F ratio as large as the one attained is very small (e.g., $<.01$), we would conclude that the two samples did not come from the same population. The use of the F ratio will be explained further in the course of the subsequent discussion.

SINGLE-FACTOR ANALYSIS OF VARIANCE

Assume that, to study the effects of background noise on students' problem-solving ability, we test three different groups of subjects, each under a different noise level, and measure the number of problems each subject solves. The means for each group are shown in Table 10.1. Does the noise level affect performance? If it does not, then, in effect, all subjects were tested under the same condition, and the three groups can be assumed to be samples from the same population. We could compare every possible pair of conditions using a t test, but this procedure would lead to large type I errors (see Chapter 5) as well as being cumbersome (particularly when there are many conditions). A single test that can be used for all groups simultaneously and will test directly whether the experimental variable has any effect is better.

TABLE 10.1. Mean Number of Problems Solved

Under Quiet Conditions	Under Slightly Noisy Conditions	Under Very Noisy Conditions
15	11	5

A first step in developing this test could be to graph the data using fictitious, smooth curves (as if the sample size were much larger), as done in Figure 10.1. For sets of samples with the same differences among group means, the distribution of scores *within* each group can range between

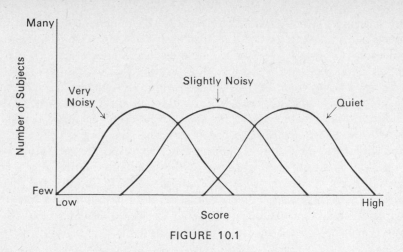

FIGURE 10.1

extremes, as shown in Figures 10.2A and 10.2B. Obviously, the differences among the group means in Figure 10.2A are relatively small compared to the spread of scores within each group, while in Figure 10.2B the differences

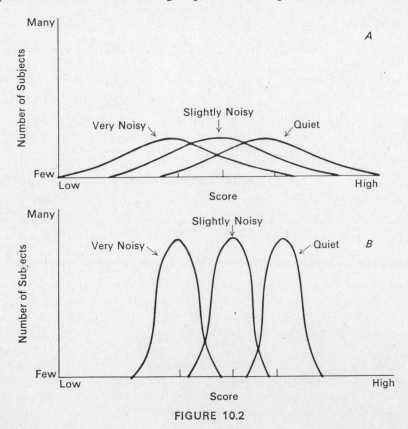

FIGURE 10.2

among means are more striking and are large compared to the very small variability within each group. In considering the t test (Chapter 9) and the White test (Chapter 4), we similarly related the difference between the two groups being compared to the spread of scores within each group.

Assume that the three samples in the example are actually drawn from the same population—that is, that noise level does not affect performance. What information is available concerning the characteristics of that population? As with the t test we are not concerned with estimating the mean of the population from the means of the samples. However, the variance of each sample (s^2) is an unbiased estimate of the population variance. In other words, the samples can provide a good picture of the variance of the population from which the samples were drawn. The formula for the t test (from Chapter 9) shows the relationship between the spread of scores (given by s^2) and the spread expected among sample means ($s_{\bar{X}}^2$):

$$s_{\bar{X}} = \frac{s}{\sqrt{N}} \qquad \text{or} \qquad s_{\bar{X}}^2 = \frac{s^2}{N} \qquad\qquad (10.1)$$

where $s^2 \rightarrow \sigma_{\text{pop}}^2$.

Thus since $s^2 = Ns_{\bar{X}}^2$, there are two ways to estimate the population variance (σ_{pop}^2):

$$s^2 \rightarrow \sigma_{\text{pop}}^2. \qquad\qquad (10.2a)$$

and

$$Ns_{\bar{X}}^2 \rightarrow \sigma_{\text{pop}}^2. \qquad\qquad (10.2b)$$

The value for $s_{\bar{X}}^2$ can be calculated from the values for the means just as the variance can be determined for any set of three scores. If the variance *among* the sample means ($s_{\bar{X}}^2$) is multiplied by the number of subjects in each sample (N), an estimate of the population variance that is separate and *independent* of the estimate derived from the average spread of scores *within* each sample will be obtained. The question now is whether or not these two estimates of the population variance are consistent—that is, are the differences *among* the means of the three groups what would be expected in view of the spread of the scores *within* each of the groups? This is exactly the same question asked with the t test. We already have at our command a technique for comparing two variance measures—the F ratio. The bare outline of the analysis-of-variance procedure presented in this section will be expanded upon in the course of analyzing the data for the example.

Sum of Squares

Although the necessary terms can be calculated directly from the mean and the variance of each of the three groups, there is a slightly different approach employing a general procedure based on original- or raw-score formulas, which is better to use in practice and is applicable to situations more complex than that used here. First, we need a measure of the total variability of all the scores in the sample. This term, which can be expressed as the sum of the squared deviations of each score about the **grand mean** ($\overline{X}_G$) of all the scores [$\sum (X - \overline{X}_G)^2$], is known as the total **sum of squares**. Each score, therefore, can be described as a deviation from the mean of all the scores ($\overline{X}_G$). Each score can also be described in terms of its deviation from the mean of its own group ($X - \overline{X}_i$) and the deviation of that mean from the grand mean ($\overline{X}_i - \overline{X}_G$); that is, $(X - \overline{X}_G) = (X - \overline{X}_i) + (\overline{X}_i - \overline{X}_G)$.

Thus we can *partition* the total sum of squares into two components, one relating to the spread of scores about the means of the individual groups (i.e., the variability within each group) and the other relating to the spread of the individual means about the grand mean (i.e., the variability among means). If all the scores in all the samples were identical, we would have a very simple, if very uninteresting, situation. But if the scores are not all identical, if there is variability among them, we may be concerned with the question: "What portion of the observed variability can be attributed to chance variability among the scores of subjects tested under a given condition (i.e., the variability about the means of the individual groups), and what portion can be attributed to the different conditions the three groups were tested under (i.e., the variability due to differences among the means of the three groups)?" In other words, the variability attributable to the differences *among* the means is compared to the variability *within* groups of subjects to determine if the differences among groups might be attributed to chance. This comparison is essentially the same as that used with the *t* test, where the numerator of the *t* ratio (DM) is the difference *between* the means, and the denominator (s_{DM}) is a function of the variability *within* each group.

It may be helpful to interpret the analysis of variance procedure in terms of a simplified presentation of the *linear model* (cf. Hays 1963). Each score (X_{ij}, or score *i* in column *j*) is the sum of three elements: the grand mean ($\overline{X}_G$); the effect of the experimental treatment on the particular group (Tr_j); and the inherent variability, or random error, for score X_{ij} (e_{ij}).

$$X_{ij} = \overline{X}_G + Tr_j + e_{ij} \qquad (10.3)$$

Consider the situation of Table 10.2 where all scores for all groups are identical. There are no treatment effects and no random error, so that all scores can be represented by $\overline{X}_G$.

$$X_{ij} = \overline{X}_G \qquad (10.4)$$

TABLE 10.2

Sample A	Sample B	Sample C
10	10	10
10	10	10
$\overline{X} = 10$	$\overline{X} = 10$	$\overline{X} = 10$
	$\overline{X}_G = 10$	

In Table 10.3 the three treatment groups differ, but there is still no variability within the groups. The scores can, therefore, be represented by the sum of $\overline{X}_G$ and the treatment effect (Tr_j):

$$X_{ij} = \overline{X}_G + Tr_j \qquad (10.5)$$

TABLE 10.3

Sample A	Sample B	Sample C
$10 - 2 = 8$	$10 + 0$	$10 + 2 = 12$
$10 - 2 = 8$	$10 + 0$	$10 + 2 = 12$
$\overline{X} = 8 = 10 - 2$	$\overline{X} = 10 = 10 + 0$	$\overline{X} = 12 = 10 + 2$
$Tr_j = -2$	$Tr_j = 0$	$Tr_j = +2$
	$\overline{X}_G = 10$	

In Table 10.4 both treatment effects and random variability (e_{ij}) affect each score. This situation requires formula (10.3):

$$X_{ij} = \overline{X}_G + Tr_j + e_{ij}$$

The random errors introduce "noise" or uncertainty into the situation, and the means of the group no longer are perfect measures of the treatment

TABLE 10.4

Sample A	Sample B	Sample C
$10 - 2 - 2 = 6$	$10 + 0 - 2 = 8$	$10 + 2 - 1 = 11$
$10 - 2 - 1 = 9$	$10 + 0 + 2 = 12$	$10 + 2 + 3 = 15$
$\overline{X} = 7.5$	$\overline{X} = 10$	$\overline{X} = 13$
	$\overline{X}_G = 10.167$	

effects. If we remove the treatment effects from Table 10.4, the random variability alone causes differences among the means of the groups, as shown in Table 10.5.

TABLE 10.5

Sample A	Sample B	Sample C
$10 - 0 - 2 = 8$	$10 + 0 - 2$	$10 + 0 - 1$
$10 - 0 + 1 = 11$	$10 + 0 + 2$	$10 + 0 + 3$
$\bar{X} = 9.5$	$\bar{X} = 10$	$\bar{X} = 11$

In the analysis of variance the task is to determine whether group differences can reasonably be attributed to random error (e_{ij}), or whether they are sufficiently large (relative to e_{ij}) to indicate they are treatment effects (Tr_j). Since e_{ij} is measured by the variability of scores within groups, differences among groups are compared to the variability within groups by analysis of variance in the same way as with the t test.

Summary Table

The component parts of the total sums of squares (SS_T) must now be translated into estimates of the population variance. The term for calculating the sum of squares for subjects within groups (SS_W) is identical to the numerator for calculating the variance, that is $\sum (X - \bar{X})^2$. Therefore, if we divide the value for SS by the appropriate df, we will obtain an estimate of the population variance. The SS for the means also gives an estimate of the population variance when divided by the appropriate df. Comparing these two estimates using an F test will permit us to decide whether the differences among the means are too large, relative to the spread of scores within each group, to assume they are due to chance.

The formulas used in calculating the values for SS for each term are shown in Table 10.6, which is in the form of a **summary table**. These formulas may look complex at first, but when numbers are used in an actual calculation (as in the worked example in Table 10.7), they prove to be basically simple and straightforward. The summary table is a standard way of depicting the calculations for analysis of variance. The heading "source" indicates the measure used. The "columns" term indicates the measure based upon the differences among means, since the columns in a table giving test data comprise the individual treatment, or experimental, groups. The "within" term refers to the variability measure based upon scores within a group. The "total" term is a measure of the variability about the grand mean. The df column shows the appropriate values for these three measures. With three experimental conditions, three columns, 1 degree of freedom is lost for the mean of these three means and, therefore, $df = C - 1 = 3 - 1 = 2$. Similarly, for the within term, based on the variability of the subjects about the means of their own groups, 1 degree of freedom is lost for each group mean and, therefore, $df = N - C$. Finally, the df for the total term involves the scores compared to the $\bar{X}_G$, so that with only one fixed value $df = N - 1$.

TABLE 10.6. Summary Table for Single-Factor Analysis

Source	df	SS	MS	F
Columns	$C - 1$	$\sum \left(\frac{T_{ci}^2}{N_{ci}} \right) - \frac{T^2}{N}$	$\frac{SS_C}{df_C}$	$\frac{MS_C}{MS_W}$
Within = Error	$N - C$	$SS_T - SS_C$	$\frac{SS_W}{df_W}$	—
Total	$N - 1$	$\sum \sum (X^2) - \frac{T^2}{N}$	—	—

$$\text{where } \frac{T^2}{N} = \frac{(\sum \sum X)^2}{N}$$

First we will deal with the term for the total SS (SS_T). The term T stands for the total of the group specified in the subscript. The double summation signs ($\sum \sum$) are instructions to sum down each column of test data and then to sum the column sums across the columns. Therefore T^2 can also be written as $(\sum \sum X)^2$ making the term for the SS_T:

$$SS_T = \sum \sum (X^2) - \frac{(\sum \sum X)^2}{N} \tag{10.6}$$

Compare formula (10.6) with the raw-score formula for s^2:

$$s^2 = \frac{\sum (X^2) - \frac{(\sum X)^2}{N}}{N - 1} \tag{10.7}$$

and (Surprise!) the formula for SS_T is identical to the numerator of the formula for s^2. Dividing SS_T by the appropriate df (i.e., $N - 1$) completes the calculation of the variance of scores about the grand mean. This is the way SS_T was originally described, but it is comforting to see it in familiar form.

Correction Term

The formula for the columns SS (SS_C) states that the total of the scores in a given column ($\sum T_{c_i}$) should be squared [$(\sum T_{c_i})^2$], and then divided by the number of scores in that column $[(\sum T_{C_i})^2/N_{C_i}]$, and that final value added to the equivalent value for each of the other columns, and T^2/N subtracted from the sum. This last value (T^2/N) is usually referred to as the **correction term** and, in the formula, is used to relate the deviations to the $\bar{X}_G$.

Mean Square

We could calculate the within SS (SS_W) directly, but this is unnecessary, since the SS_T has been partitioned into only two parts, and we have calculated one (SS_C); the other part can then easily be obtained by subtraction.

Dividing the value for SS_C by the appropriate df provides an estimat· of the population variance. Dividing the SS term by df provides a measure of the average SS for each df and is called the **mean square** (MS).

Error Term

We are interested only in the situation where the difference among means (MS_C) is large relative to the variability of scores within each group (MS_W). More specifically, we are interested in the situation where the estimate of the population variance based upon differences among means (MS_C) is large relative to the population variance estimate based upon the spread of scores within each sample (MS_W). Therefore the only F ratio tested is MS_C/MS_W. We are rarely interested in the unlikely situation where MS_C is smaller than MS_W. The MS_W is a measure of the variance among subjects treated alike (i.e., subjects within groups); it is, therefore, the best measure of the inherent variability among the subjects (i.e., the variability not due to the experimental manipulation). Variability can be considered as a measure of unpredictability or error (the term "error" being used as in "error of estimate" about regression lines), and the reference term, or denominator, of the F ratio is usually referred to as the **error term**.

The F ratio can also be considered in terms of the linear model for expressing each score. The MS_W is a measure of the random error (e_{ij}), while the MS_c is a measure of the *two* factors that contribute to column differences—error (e_{ij}) *and* treatment effects (Tr_j). This description of the F ratio is summarized as follows:

$$F = \frac{MS_C}{MS_W} = \frac{e_{ij} + Tr_j}{e_{ij}} \tag{10.8}$$

If $Tr_j = 0$

then $F = \dfrac{e_{ij}}{e_{ij}} = 1$

TABLE 10.7

Number of Problems Solved Under:

Quiet Conditions	Slightly Noisy Conditions	Very Noisy Conditions
18	14	8
17	12	6
15	9	5
15	9	4
10		2
$T = 75$	$T = 44$	$T = 25$
$N = 5$	$N = 4$	$N = 5$

Summary Table

Source	df	SS	MS	F	p	r_m
Columns	$3 - 1$	$\frac{(75)^2}{5} + \frac{(44)^2}{4} + \frac{(25)^2}{5} - \frac{T^2}{N} = 1{,}734 - 1{,}481.14 = 252.86$	126.43	18.30	<.001	>.80
Within	$14 - 3$	$SS_T - SS_C = 328.86 - 252.86 = 76$	$\frac{76}{11} = 6.91$			
Total	$14 - 1$	$(18)^2 + (17)^2 + \cdots + (2)^2 - \frac{T^2}{N} = 1{,}810 - 1{,}481.14 = 328.86$				

$$\frac{T^2}{N} = \frac{(18 + 17 + 15 + \cdots + 4 + 2)^2}{14} = \frac{(144)^2}{14} = 1{,}481.14$$

If the F ratio is significantly larger than 1.0, then we conclude that the treatment, or the experimental variable (Tr_j), actually has an effect; that is, $Tr_j > 0$.

For the noise-level example of Table 10.1 the individual scores and summary table are given in Table 10.7, with the probability values for the F ratio taken from Table E and the r_m value taken from Table B. There are large $(r_m > .80)$ and significant $(p < .001)$ differences in the subjects' performance under the three noise levels, with the most problems being solved under the quiet condition and fewest problems being solved under the very noisy condition. This conclusion is the same as when the Kruskal-Wallis test was used for these data (Table 5.1), except that the extra information available in using the actual score values instead of the ranks (with the $\bar{X}$ and the s being suitable for these data) leads to a slightly smaller value for p and a larger value for r_m.

Once again, to compare two of these conditions we would use the appropriate two-sample test—in this case the t test. We have to be cautious when doing multiple two-sample tests on the basis of a multi-sample comparison, since we are increasing the opportunity to make type I errors, that is, to mistakenly reject the null hypothesis (cf. Kirk 1968). In this study, however, we were primarily interested in a comparison of the subjects' performance under the slightly noisy and the very noisy conditions, and the t test is calculated in the following fashion:

$$\bar{X}_{\text{S.Noisy}} = 11 \qquad \bar{X}_{\text{V.Noisy}} = 5$$

$$DM = 6 \qquad s_{DM} = 1.5629$$

$$t = \frac{\bar{X}_1 - \bar{X}_2}{s_{DM}} = \frac{11 - 5}{1.5629} = 3.8390 \qquad (10.9)$$

$$p < .01 \qquad r_m > .80$$

There are large $(r_m > .80)$ and significant $(p < .01)$ differences in the scores under the two conditions, the better scores occurring under the slightly noisy condition. This is the same conclusion as for the White test which was done on this specific comparison after the Kruskal-Wallis test showed significant differences across all three groups.

Take a look now at the analysis of variance of the data of these same two groups given in Table 10.8. The values for r_m and p are the same as for the t test, and a closer look reveals that the calculated value for the F ratio (14.7370) is equal to the square of the t ratio (3.8390). When there are only two groups (df for the numerator of the F ratio = 1), then $t^2 = F$. A check of Table E for the F ratio reveals that the terms in the column for $df = 1$ for the numerator are the squares of the terms in Table A for the t ratio. This means that the t test is a special version of the F test for two samples and has somewhat simplified calculations. The conclusions and interpretations are

TABLE 10.8

Source	df	SS	MS	F	p	r_m
Columns	2 − 1	$\frac{(44)^2}{4} + \frac{(25)^2}{5} - \frac{T^2}{N} = 609 - \frac{T^2}{N} = 80.00$	80.00	14.7370	<.01	>.80
Within	9 − 2	$118 - 80 = 38$	$\frac{38}{7} = 5.4285$			
Total	9 − 1	$(14)^2 + (12)^2 + \cdots + (2)^2 - \frac{T^2}{N} = 647 - \frac{T^2}{N} = 118.00$				

$$\frac{T^2}{N} = \frac{(14 + 12 + \cdots + 4 + 2)^2}{9} = \frac{(69)^2}{9} = 529.00$$

necessarily the same for the same data whether the *t* test or analysis of variance test is used.

Helpful hint in calculation: It is impossible to get a negative value for *SS* if the calculations are done correctly. If you get a negative value, there is a mistake somewhere—even if it is not *your* fault—and you have to correct it before you can proceed. A mistake in calculation that leads to a negative *SS* is (in a sense) a lucky event, since the error is conspicuous and cannot be overlooked. Errors are easy to make when working an analysis of variance, and all calculations should be checked. It is poor policy to check only those outcomes that do not agree with your preconceived notions. When you check calculations, check the ones giving outcomes you are most happy with. You will check the others in any case.

SINGLE-FACTOR ANALYSIS OF VARIANCE

Formula

Source	df	SS	MS	F
Columns	$C - 1$	$\sum \left(\dfrac{T_{C_i}^{2}}{N_{C_i}} \right) - \dfrac{T^2}{N}$	$\dfrac{SS_C}{df_C}$	$\dfrac{MS_C}{MS_W}$
Within = Error	$N - C$	$SS_T - SS_C$	$\dfrac{SS_W}{df_W}$	
Total	$N - 1$	$\sum \sum (X^2) - \dfrac{T^2}{N}$		

$$\text{where } \frac{T^2}{N} = \frac{(\sum \sum X)^2}{N}$$

Table E provides p values based on F, df_C (as df_N) and df_W (as df_D).

Table C provides r_m values based on the F ratio, df_C (as df_N) and df_W (as df_D).

Restrictions The scores within each sample should be close to normally distributed with the mean and standard deviation being suitable measures. The variance of the samples should be roughly equal. The effect of skewness, or unequal variance, decreases as the sample sizes increase. An alternative test when the scores are skewed and the sample sizes are reasonably small is the Kruskal-Wallis test.

Procedure
1. Obtain the values of $\sum X$ ($= T_C$) and $\sum (X^2)$ for each sample.
2. Calculate T^2/N by summing all the $\sum X$ terms from step 1 to obtain the sum of all the scores. Square this sum and divide by the total number of scores (N).

3. Calculate the total SS term (SS_T) by adding the $\sum (X^2)$ terms obtained in step 1 and subtract T^2/N.
4. Calculate the columns SS term (SS_C) by squaring each T_C ($= \sum X$) term from step 1 and dividing that squared term by the number of scores in that group. Add the resulting values and subtract T^2/N.
5. Calculate the within, or error, term (SS_W) by subtracting SS_C from SS_T.
6. Calculate df for each term.
7. Divide SS_C by df_C to obtain MS_C.
   Divide SS_W by df_W to obtain MS_W.
8. Divide (MS_C) by (MS_W) to obtain the F ratio.
9. Use Table E to determine the p values based on F, df_C (as df_N), and df_W as (df_D).
10. Use Table C to determine the r_m value based on F, df_N, and df_D.

Example

Three groups of subjects are given two-minute tests including identical sets of math problems. The members of the first group are told that the problems are being pretested and they should try to do well (low-drive condition). The members of the second group are offered five cents for each correctly solved problem (medium-drive condition). The third group is told that performance on the test is highly correlated with intelligence and social success (high-drive condition).

Number of Problems Solved		
Low-Drive Condition	Medium-Drive Condition	High-Drive Condition
1	3	6
2	4	8
2	5	8
4	7	9
$\sum X_L$	$\sum X_M$	$\sum X_H$

$$\frac{T^2}{N} = \frac{(\sum X_L + \sum X_M + \sum X_H)^2}{12}$$

$$= \frac{(1 + 2 + 2 + 4 + \cdots + 6 + 8 + 8 + 9)^2}{12}$$

$$= \frac{(59)^2}{12} = 290.0833$$

$$\text{total } SS = [\sum (X^2)_L + \sum (X^2)_M + \sum (X^2)_H] - \frac{T^2}{N}$$

$$= (1^2 + 2^2 + 2^2 + 4^2 + \cdots + 6^2$$
$$+ 8^2 + 8^2 + 9^2) - \frac{T^2}{N}$$

$$= 369.00 - 290.0833 = 78.9167$$

$$\text{columns } SS = \left[\frac{(\sum X_L)^2}{N_L} + \frac{(\sum X_M)^2}{N_M} + \frac{(\sum X_H)^2}{N_H}\right] - \frac{T^2}{N}$$

$$= \frac{(1 + 2 + 2 + 4)^2}{4} + \cdots$$
$$+ \frac{(6 + 8 + 8 + 9)^2}{4} - \frac{T^2}{N}$$

$$= 20.25 + 90.25 + 240.25 - \frac{T^2}{N}$$

$$= 350.75 - \frac{T^2}{N} = 60.6667$$

$$\text{within } SS = SS_T - SS_C = 78.9167 - 60.6667$$
$$= 18.2500$$

Source	df	SS	MS	F	p	r_m
Columns	3−1=2	60.6667	30.3333	14.96	<.01	>.80
Within = Error	12−3=9	18.2500	2.0278			
Total	12−1=11	78.9167				

Conclusion

There are significant ($p < .01$) and large differences ($r_m > .80$) in performance under the three drive conditions. The most problems were solved by those working under the high-drive condition and the fewest by those in the low-drive group.

SUMMARY

1. The F ratio is a ratio of variances, that is, $s_1{}^2/s_2{}^2$. The mean of the sampling distribution for the F ratio, which is the value expected by chance, is 1.0.

2. The *single-factor analysis of variance* is an extension of the t test to three or more conditions and is analogous to the Kruskal-Wallis test.

3. The deviation of each score from the *grand mean*, or mean of all the scores, can be described in terms of the deviation of that score from the mean of the sample plus the deviation of the sample mean from the grand mean. The deviations of scores from sample means and of sample means from the grand mean are squared and summed to yield the *sum of squares* for scores within samples and for means about the grand mean.

4. The *summary table* is a convenient way of depicting the calculations involved in an analysis of variance.

5. The *correction term*, or $(\sum \sum X)^2/N$, is used in the calculation of sums of squares with a raw-score formula.

6. The *mean square* is the sum of squares divided by the appropriate df and provides an independent estimate of the population variance.

7. The mean square for subjects treated alike, or the variability not attributable to the experimental manipulations, is the best approximation or estimate of the "true" population variance, or error, and this term is called the *error term*. The error term is used as the denominator of the F ratio to test the other term(s) in the analysis.

chapter eleven

TWO-FACTOR ANALYSIS OF VARIANCE

In Chapter 10 the total variability among scores was analyzed to determine whether the samples could reasonably have been drawn from the same population. An example involved testing the effects of three different levels of background noise on students' problem-solving ability. We now wonder if the effect of noise is the same when the subjects are doing easy problems as when they are doing hard ones. How can this possibility be tested? The obvious way is to test the subjects' ability to solve easy and hard problems under quiet conditions, easy and hard problems under slightly noisy conditions, and finally easy and hard problems under very noisy conditions. The problem is to analyze the data from this experiment. A t test between the number of easy and hard problems solved under each of the three noise conditions would give an idea of the differences between solving easy and hard problems, but the overall effect might not be clear. What about the three noise conditions? We could combine the scores for solving easy and hard problems under each noise condition and then do a single-factor analysis of variance to find any overall differences in performance under the noise conditions. This is clearly unsatisfactory, since any differences between the solving of easy and hard problems would increase the variability within the groups working under each noise condition and obscure differences among the groups by making the error term (MS_W) large compared to MS_C. An extension of the single-factor analysis of variance provides a solution to this problem. A **two-factor analysis of variance** can be applied to data from an experiment having two variables and separate groups of subjects for each combination of these variables (e.g., 2 task levels × 3 noise levels = 6 groups). This design is summarized in Figure 11.1.

173

Background Noise Level

		Quiet	Slightly Noisy	Very Noisy
Problem Difficulty	Easy	Scores A B C	Scores D E F	Scores G H I
	Hard	Scores J K L	Scores M N O	Scores P Q R

FIGURE 11.1

The experimental situation in which each level of one variable, or factor, is combined with each level of another variable, or factor, is called a **factorial design**. By partitioning the total variability into the portion contributed by one variable (e.g., task difficulty), we can test differences in performance under different noise-level conditions as if there were no task variable, and we can test the difference between solving easy and difficult problems as if there were no noise variable. However, the initial concern of this chapter is with whether or not the effect of increasing the noise level might vary with different levels of task difficulty. Consider the data from the experiment diagramed in Figure 11.2.

Background Noise Level

		Quiet		Slightly Noisy		Very Noisy			
Problem Difficulty	Easy	17 15 13	45 638	15 13 11	39 515	12 10 8	30 308	$\leftarrow \sum X$ $\leftarrow \sum X^2$	$T_E = 114$
	Hard	15 14 13	42 590	10 8 6	24 200	4 3 2	9 29		$T_H = 75$

$87 = T_Q$ $63 = T_{SN}$ $39 = T_{VN}$

FIGURE 11.2: Number of Problems Solved by Each Subject

Clearly, performance tends to be better under quieter than noisier conditions, regardless of task difficulty. In addition, the number of easy problems solved is greater than the number of hard problems solved, regardless of noise level. This last finding is essentially uninformative, beyond showing that we actually used "easy problems" and "hard problems." The effect of noise level can be obtained by averaging the scores for both task difficulties at each noise level, and the effect of task difficulty by averaging scores across the three noise levels for each task level, as shown in Figure 11.3. But there is something else: Under the quiet condition the difference between the scores for the easy and hard tasks is relatively small; as the noise level increases,

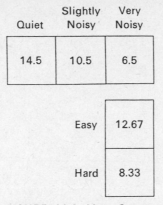

FIGURE 11.3: Mean Correct for Each Factor (Averaged Across the Other Variable)

the difference between the scores for the two types of tasks increases such that the solving of the easy problems shows a relatively small decline while solution of the hard problems shows a relatively sharp decline. This particular effect is not a function of noise level alone nor of task difficulty alone but rather is the result of the combination, or **interaction**, of the two factors.

Interaction

Interaction is the last major concept included in this text and is the primary reason for discussing the two-way analysis of variance. Once interaction is explained, the two-way analysis-of-variance procedure will be readily understood as an extension of single-factor analysis.

Two variables are said to interact when the effects of one variable are at least partially determined by the level of the other variable. In our example the very noisy background barely interfered with performance on the easy problems (compared to the quiet background) but did severely disrupt performance on the hard problems. Conversely, the hard problems were solved only slightly less often than the easy problems under the quiet condition, but as the noise level increased, solution of the hard problems became increasingly less frequent.

When the data for a two-variable study are graphed, interaction is evident when the curves for the two variables are not *parallel*, that is, when there is a *difference in slope* among the curves, or lines, on the graph. This is the visual test for interaction; the statistical test for interaction indicates the likelihood that the difference in slope is due to chance.

Assume that the data in our example is as shown in Figure 11.4. Here the two lines are parallel, showing that there is no interaction. Performance is affected by noise level and tends to be worse with higher noise levels than with lower noise levels. There is no need to make any reference to the

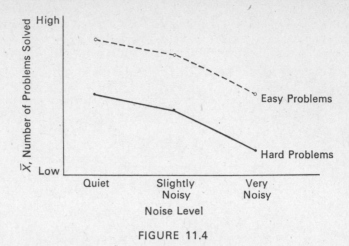

FIGURE 11.4

difficulty of the task, since the conclusion holds equally for both easy and hard problems. Similarly, we could describe the effect of task difficulty by saying that more easy problems are solved than hard problems. Here again, we need not make any reference to background noise level, since the conclusion holds for all the noise levels.

To describe the effect of background noise level on the problem-solving ability shown in Figure 11.2, we first have to indicate whether we are talking about easy problems or hard problems. In describing the difference between solving easy and hard problems, we must give a different statement for each level of background noise. We cannot simply describe the effect of either of the variables without taking into account the other variable, since each variable affects or acts upon the other. In other words, there is an interaction between these two variables.

A very clear example of interaction occurs in a situation in which both Democrats and Republicans are tested on the degree of their agreement with speeches of the most recent Democratic and Republican presidential candidates (D.C. and R.C.). A reasonable outcome is shown in Figure 11.5. When the degree of agreement is averaged across D.C.'s and R.C.'s speeches, it is identical for Democratic and Republican subjects. Similarly, when the level of agreement is averaged across the Democratic and Republican groups, it is the same for D.C.'s and R.C.'s speeches. In other words, neither of the experimental variables (party affiliation or candidate affiliation) seems to make a difference. Yet there is certainly a very large effect. The difference between the two slopes is quite compelling, and the only important effect is the interaction between the two variables. The appropriate conclusion is that the overall agreement is the same for D.C.'s and R.C.'s speeches and that Democratic and Republican subjects are equally likely to approve given speeches. The interaction, however, reflects that Democratic subjects tend to

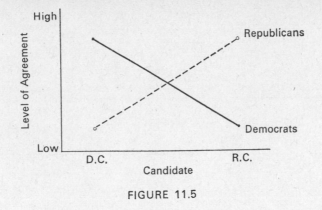

FIGURE 11.5

agree with D.C.'s speeches and disagree with R.C.'s speeches and the reverse is true for Republican subjects.

Knowledge about the interaction between two variables is not readily obtained when single-variable experiments are conducted. In terms of our original example, we would have no knowledge of the interaction after testing solution of hard and easy problems under only quiet or only noisy conditions. There must be at least two levels of both variables before a picture of any possible interaction can be obtained. The next problem is that of testing for the presence of an interaction.

In the two-way analysis of variance the portion of the total variability attributable to each of the two variables, or factors, is determined in the same way as in the single-factor analysis. There is also a component that represents the variability among subjects within groups, or treated alike (the MS_w), which is used as the best estimate of the population variance (i.e., the error term) and as the denominator of the F ratio. In addition to these three terms, the division, or partition, of the total variability (SS_T) includes a term that refers to the variability attributable to the difference in slope for the various groups—that is, attributable to interaction. The test for interaction,

		Columns	
		A	B	
Rows	1	$\bar{X} = 12$ $\bar{X}_{ij} = 9 + 5 - 2$	$\bar{X} = 16$ $\bar{X}_{ij} = 9 + 5 + 2$	$\bar{X} = 14 = 9 + 5$ $Tr_{R1} = +5$
	2	$\bar{X} = 2$ $\bar{X}_{ij} = 9 - 5 - 2$	$\bar{X} = 6$ $\bar{X}_{ij} = 9 - 5 + 2$	$\bar{X} = 4 = 9 - 5$ $Tr_{R2} = -5$
		$\bar{X} = 7 = 9 - 2$ $Tr_{C_A} = -2$	$\bar{X} = 11 = 9 + 2$ $Tr_{C_B} = +2$	$\bar{X}_G = 9$

$$X_{ijk} = \bar{X}_G + Tr_{R_i} + Tr_{C_j} \, (+ e_{ijk})$$

FIGURE 11.6

Columns

	A	B	
1	$\bar{X} = 11$ $I_{ij} = -1$ $\bar{X}_{ij} = 9 + 5 - 2 - 1$	$\bar{X} = 17$ $I_{ij} = +1$ $\bar{X}_{ij} = 9 + 5 + 2 + 1$	$\bar{X} = 14 = 9 + 5$ $Tr_{R_1} = +5$
2	$\bar{X} = 3$ $I_{ij} = +1$ $\bar{X}_{ij} = 9 - 5 - 2 + 1$	$\bar{X} = 5$ $I_{ij} = -1$ $\bar{X}_{ij} = 9 - 5 + 2 - 1$	$\bar{X} = 4 = 9 - 5$ $Tr_{R_2} = -5$

Rows

$\bar{X} = 7 = 9 - 2$ $\bar{X} = 11 = 9 + 2$

$Tr_{C_A} = -2$ $Tr_{C_B} = +2$ $\bar{X}_G = 9$

$$X_{ijk} = \bar{X}_G + Tr_{R_i} + Tr_{C_j} + I_{ij}\,(+e_{ijk})$$

FIGURE 11.7

therefore, is essentially the same as for the separate variables, and the calculation is no more difficult.

Interpreting the factorial design in terms of the linear model may help clarify the meaning of the column and row effects and interaction. In the example given in Figure 11.6 there are only row and column treatment effects (Tr_{R_i} and Tr_{C_j} for row i and for column j). To simplify matters, just the means for each group are shown as if there were no random error in the individual scores within each group; that is, e_{ijk} (for error of score k in row i, column j) $= 0$. In actual data random errors would always be present. Note that each group mean can be described as the sum of $\bar{X}_G$ + row i effect + column j effect.

In Figure 11.7 the mean values for each group cannot be obtained from the simple addition of $\bar{X}_G$ with the row and column effects, even though the overall values of Tr_{R_i} and Tr_{C_j} are the same as in Figure 11.6. Each group shows the effect of the interaction of the row and column variables in addition to the overall row and column treatment effects. In the example the absolute value of the interaction (I_{ij}) is the same for each group, but this would rarely occur in actual practice.

In the two-way analysis of variance an F ratio is used to test each of the possible effects (row, column, and interaction) separately against the term for random error.

Calculation of Two-Way Analysis of Variance

The analysis follows the same procedure as used for the single-factor design. An intermediate term (cells) is included, which serves to simplify and facilitate the calculations. The computational formulas are shown in Table 11.1.

TABLE 11.1. Summary Table for Two-Way Analysis of Variance

Source	df	SS	MS	F
Columns	$C - 1$	$\sum \left(\dfrac{T_{Ci}^2}{N_{Ci}} \right) - \dfrac{T^2}{N}$	$\dfrac{SS_C}{df_C}$	$\dfrac{MS_C}{MS_W}$
Rows	$R - 1$	$\sum \left(\dfrac{T_{R_j}^2}{N_{R_j}} \right) - \dfrac{T^2}{N}$	$\dfrac{SS_R}{df_R}$	$\dfrac{MS_R}{MS_W}$
(Cells)	—	$\sum \sum \left(\dfrac{T_{C_iR_j}^2}{N_{C_iR_j}} \right) - \dfrac{T^2}{N}$	—	—
$C \times R$ Interaction	$(C - 1)(R - 1)$	$SS_{\text{cells}} - SS_C - SS_R$	$\dfrac{SS_{CR}}{df_{CR}}$	$\dfrac{MS_{CR}}{MS_W}$
Within	$N - RC$	$SS_T - SS_{\text{cells}}$	$\dfrac{SS_W}{df_W}$	—
Total	$N - 1$	$\sum \sum (X^2) - \dfrac{T^2}{N}$	—	—

Comparing Table 11.1 to the formulas for the single-factor analysis of Table 10.6, we can see that the total SS is identical, and the formulas for "columns" (the variable graphed on the abscissa, which reads up and down like columns) and for "rows" (which is graphed on the ordinate and reads across) are identical to the formulas for the columns in the single-factor analysis. In both cases the scores in each column or row are added to obtain the total for that column or row (T). This value is then squared and divided by the number of scores in that column (N). These steps are repeated for each of the columns (or rows). The values are added together, and the correction term is then subtracted from the final sum. The cells SS are a measure of the variability among the cells in the analysis. Each cell contains a group of subjects tested under one of the combinations of experimental conditions, or factors. In the example there are three noise levels and two levels of task difficulty giving a 3×2 factorial design with $3 \times 2 = 6$ cells. The SS term for cells is calculated exactly as the columns and rows SS, except the scores are summed only within each cell; that sum is squared and divided by the number of scores within that cell; this process is repeated for all the cells; and finally, T^2/N is subtracted from the final sum. The SS_{cells} is used to permit the calculation of the interaction term and the within term by subtraction. The within, or error, term is the average variance of the scores *within* each of the cells—in other words, the average variance of the scores among subjects who are treated alike. The example would give the following calculations:

$$SS_C = \frac{(45 + 42)^2}{6} + \frac{(39 + 24)^2}{6} + \frac{(30 + 9)^2}{6} - \frac{T^2}{N} \qquad (11.1)$$

$$= 192.00$$

$$SS_R = \frac{(45 + 39 + 30)^2}{9} + \frac{(42 + 24 + 9)^2}{9} - \frac{T^2}{N} \tag{11.2}$$

$$= 84.50$$

$$\text{cells} = \frac{(45)^2}{3} + \frac{(39)^2}{3} + \frac{(30)^2}{3} + \frac{(42)^2}{3} + \frac{(24)^2}{3} + \frac{(9)^2}{3} - \frac{T^2}{N} \tag{11.3}$$

$$= 304.50$$

$$\text{total} = 2{,}325 - \frac{T^2}{N} = 340.50 \tag{11.4}$$

$$\frac{T^2}{N} = \frac{(189)^2}{18} = 1{,}984.50 \tag{11.5}$$

The values for df also follow the single-factor analysis, and once again the degrees of freedom for all the terms add up to the df for total. As before the MS for each term is equal to the SS divided by the df, and the F ratios for the three tests use the MS_W as the denominator. We then end up with the summary table shown as Table 11.2.

TABLE 11.2

Source	SS	df	MS	F	p	r_m
Columns	192.00	2	96.00	32.00	<.001	>.78
Rows	84.50	1	84.50	28.17	<.001	>.78
(Cells)	304.50	—	—	—	—	—
$C \times R$	28.00	2	14.00	4.67	<.05	>.53
Within	36.00	12	3.00	—	—	—
Total	340.50	17	—	—	—	—

The interpretation of the data is based on the summary table. There are actually three separate statistical tests with three separate conclusions, one each for columns, rows, and interaction. Any one of these three may show large or small, significant or nonsignificant, effects independently of the others. Although the three tests are calculated simultaneously and are based on the same data, they are essentially separate tests and are interpreted accordingly. The columns effect (noise level) is large and significant ($r_m > .78$, $p < .001$), with fewer problems solved as the noise level increases. The rows effect (problem difficulty) is large and significant ($r_m > .78$, $p < .001$), with the hard problems tending to be solved less often than the easy ones. The interaction is a moderate ($r_m > .53$) and significant ($p < .05$) effect, such that increases in noise level are associated with a larger decrease in performance for the hard task than for the easy task. The interpretation of the interaction can be rephrased by saying that the difference between solution of the hard and easy tasks is small under the quiet condition but becomes larger as the noise level increases. An equivalent interpretation is that increases in noise level have relatively small effects when the task is easy but cause a large decline in performance when the task is difficult. These three separate conclusions (for rows, columns, and interaction) are the same as those we obtained on the basis of a visual examination of the data. The analysis of variance provides objective p and r_m values to help in attributing importance to the separate effects.

Look again at Summary Table 11.1 giving the calculations for the two-way analysis. If there were only one experimental variable instead of two, the situation would be somewhat different. First of all, there would be columns variable but no rows variable, so the rows variable would not appear in the summary table. Similarly, if there is no rows variable, there are no cells, eliminating the line for cells from the table. Finally, if there is only one variable, there is no opportunity for an interaction. What is left is a summary table consisting of a measure of total variability that is divided into two components, one for columns and one for within groups. This is identical to the summary table for the single-factor analysis (Table 10.6). In other words, the single-factor analysis is a special version of the two-way analysis for when there is only one variable. We can also look at the two-way analysis of variance as an extension of the single-factor analysis, which copes with the two-factor situation by dividing the total sum of squares into components appropriate to each of the two factors, the interaction between the factors, and the variability *within* cells.

Treatments-by-Subjects Design and Residual Variance

The data given in Chapter 5 for a repeated-measures design and analyzed by the Friedman test is repeated in Table 11.3. We can now treat these data in terms of the actual scores for each subject rather than the ranks of the scores. Table 11.3 has the form of a two-way analysis-of-variance situation. The experimental, or independent, variable is one of the factors, and the individual subjects compose the other factor. The two-way analysis requires three columns (one for each experimental condition) and four rows (one for each subject). Set up in this fashion, the analysis follows the previous example except for a few minor changes. The cell in the two-way analysis is the intersection of a given row and a given column, and the term SS_W is the measure of the average variance among the scores within each cell. But in the present situation there is only one score per cell, and obviously, there can be no variance of the scores within a cell about the mean of that cell. Therefore, in the summary table there is no SS_W term to use as an error term for the denominator of the F ratio. The appropriate procedure in this case is to use the remaining, or residual, variance as the reference term for the population variance. The **residual variance** in this analysis is the variance left over, or not attributable to the columns and rows effects, and is the best available estimate of the population variance. The SS_W term in the single-factor analysis may similarly be considered as the residual variance. Therefore, the total sum of squares is divided into three terms: a component for columns (the experimental conditions), a component for rows (relating to differences among the subjects), and the error term.

TABLE 11.3

Subject	Quiet Condition	Slightly Noisy Condition	Very Noisy Condition
a	22	16	14
b	15	12	8
c	10	6	7
d	8	5	3

The calculations for the data are shown in Table 11.4. There are large ($r_m > .83$) and significant ($p < .01$) differences in the performance of the subjects working under the different noise levels, with the best performance under quiet and the worst under noisy conditions. The conclusion is the same as that arrived at with the Friedman test, but the differences are more pronounced when the data are used directly rather than in the form of ranks.

It should now be somewhat clearer how related-samples tests gain their efficiency. Suppose we used independent groups instead of testing each subject three times. The analysis of the example data would give the same value for the SS_T and the SS_C. The reference term (the error term), or SS_W, would be obtained by subtracting the column SS from the total SS, which would result in a fairly large value ($MS = 28.83$) for the error term, and there would be a small and nonsignificant difference among the columns. The related-samples analysis is exactly the same as the single-factor analysis, except that in the related-samples analysis the portion of the variability attributable to the differences among subjects is effectively subtracted from the error term. With the smaller error term, differences among the columns can be tested for without having the effect obscured by differences among the subjects. Essentially, in the related-samples test we are using as the reference term the variability of scores for each subject (or within each matched set of subjects) rather than the variability among all the subjects in a group (as would be the case in the independent-groups design). Using related samples is, therefore, more efficient than using independent groups, since the elimination of subject differences from the error term makes it possible to detect smaller differences among the experimental conditions. To summarize, the formulas for the calculation of SS_e for both analyses are as follows:

independent groups:

$$SS_e = SS_T - SS_C \tag{11.6}$$

related samples:

$$SS_e = SS_T - SS_C - SS_{\text{Subj.}} \tag{11.7}$$

TABLE 11.4. Treatments-by-Subjects Design (Related Samples)

Source	df	SS	MS	F
Columns	$C - 1$	$\sum \left(\dfrac{T_{C_i}^2}{N_{C_i}}\right) - \dfrac{T^2}{N}$	$\dfrac{SS_C}{df_C}$	$\dfrac{MS_C}{MS_e}$
Rows (Subjects)	$R - 1$	$\sum \left(\dfrac{T_{R_j}^2}{N_{R_j}}\right) - \dfrac{T^2}{N}$	$\dfrac{SS_R}{df_R}$	$\dfrac{MS_R}{MS_e}$
Residual Error	$(C - 1)(R - 1)$	$SS_T - SS_C - SS_R$	$\dfrac{SS_e}{df_e}$	—
Total	$CR - 1$	$\sum \sum (X^2) - \dfrac{T^2}{N}$	—	—

Subjects (rows)	Quiet Condition	Slightly Noisy Condition	Very Noisy Condition	$\sum$ row
a	22	16	14	52
b	15	12	8	35
c	10	6	7	23
d	8	5	3	16
	$\sum X = 55$	$\sum X = 39$	$\sum X = 32$	$126 = T$
	$\sum (X^2) = 873$	$\sum (X^2) = 461$	$\sum (X^2) = 318$	

	df	SS	MS	F	p	r_m
Columns	2	$\dfrac{(55)^2}{4} + \ldots - \dfrac{T^2}{N}$ $= 69.50$	34.75	18.68	$<.01$	$>.83$
Rows	3	$\dfrac{(52)^2}{3} + \ldots - \dfrac{T^2}{N}$ $= 248.3333$	82.78	44.50	$<.001$	$>.92$
Error	6	$= 11.17$	1.86	—	—	—
Total		$= 873 + 461 + 318 - \dfrac{T^2}{N} = 329$	—	—	—	—

$$\frac{T^2}{N} = \frac{(126)^2}{12} = 1323.00$$

When the $SS_{\text{Subj.}}$ can be measured and subtracted from the SS_e term, a given difference among columns (SS_C) is more likely to result in a large and significant F ratio.

As you may suspect from the discussion of single-factor analysis, when there are only two experimental conditions, the treatments-by-subjects analysis is identical to the t test for related samples. The difference between the independent and related-samples t test can be illustrated by reference to the equivalent analysis of variance for the related-samples design; the error term (i.e., $s_{\bar{D}}$) is reduced by taking out that variability attributable to subject differences. This is accomplished directly in the two-sample case by using only the difference scores and ignoring the overall level of the subjects' performance. The final outcome is the same whether the subject differences are removed before or during the analysis (either by an analysis of variance or use of a formula based on the correlation measure). At this point, the analysis-of-variance procedure may seem preferable to the t test, since it also permits testing for differences among subjects. Testing subject differences, however, is not much of an advantage, since we nearly always find significant differences among the subjects. (People or animals really do differ from one another—terms like "birds of a feather" or "peas in a pod" notwithstanding.) In fact, the choice of a related-samples design involves the assumption, in advance, that the differences among subjects are sufficiently large to make using matched samples or repeated measures on the same subject worthwhile. The t test for related samples, therefore, involves no important loss of information, and the very simple calculations required make the t test a convenient alternative to the analysis of variance.

Since the t test and the related-samples analysis of variance are so closely related, all of the problems of using related samples already discussed hold for both tests. Specifically, repeated measures on the same subject involve possible sequential and time-dependent effects, these factors becoming increasingly serious as the number of experimental conditions increases. Similarly, obtaining matched subjects becomes increasingly difficult as the number of subjects in each set increases. Given these problems, it is often simpler to use more subjects (though, perhaps, less total effort) in an independent-groups design. One way to avoid some of the difficulties in using related samples to deal with large subject differences is discussed in the next section.

Treatments-by-Levels Design

The **treatments-by-levels** experimental design is a convenient and efficient technique, which statistically controls for large differences among subjects by removing much of the variability attributable to these differences from the error term. At the same time this test provides a measure of the subject differences and of the interaction of these differences with the independent variable. As we have seen, this interaction cannot be measured in the simple treatments-by-subjects design. The treatments-by-levels design, a version of the independent-groups design, avoids or minimizes the many problems associated with related-samples designs. This analysis also has the advantage of requiring absolutely no new materials, and the calculations are exactly the same as for the two-way analysis of variance.

First consider a situation in which students of widely varying mathematical ability are randomly selected, and the effects of background noise on their solution of math problems are measured. With a very large variability within the groups (SS_W) (i.e., large differences in basic ability among the subjects), differences due to the independent variable (SS_C) are difficult to detect. If each subject is tested under each of the conditions in a treatments-by-subjects design, there is no way to measure the interaction between the ability of the subject and noise level. In addition, there is also a considerable problem with sequential effects—practice and boredom, particularly.

Background Noise Level

		Quiet	Slightly Noisy	Very Noisy
	Average	17	15	12
		15	13	10
		13	11	8
Ability				
	Pathetic	15	10	4
		14	8	3
		13	6	2

FIGURE 11.8 : Number of Problems Solved by Each Subject

On the basis of a pretest or the previous semester's grades in mathematics, the subjects can be suitably divided into two groups in terms of their ability—the adequate and the pathetic. In practice the use of three equal size groups instead of two would probably be better; but for purposes of the example two groups will suffice. The variability within each of these subgroups, or levels of ability, will be considerably less than within a randomly selected group of subjects. The levels are based on a matching procedure, and by the use of a more accurate criterion (pretest) measure or the use of more groups, the variability within each level can be decreased further. The lower the variability within levels, the easier it is to detect the effects of the experimental variable, but the greater the cost, in time and effort, to do the matching. In general, the matching within levels will not be as precise as that used within sets in a related- or matched-samples design. The data for the experiment are shown in Figure 11.8.

If these data look familiar, it is because they are identical to the data in Figure 11.2. In other words, this situation is identical to the one previously analyzed by a two-way analysis of variance, but now the subjects are divided in two groups that are relatively homogeneous with regard to a relevant variable. The summary table is the same as shown in Table 11.2. We can base our conclusions on the F values already obtained. For columns there are large ($r_m > .78$) and significant ($p < .001$) differences among noise conditions such that the best performance occurred under the quiet condition and the worst performance under the very noisy condition. For rows there are large ($r_m > .78$) and significant ($p < .001$) differences between the adequate and pathetic students such that the adequate students tended to solve more problems than the pathetic students. For the interaction term we find a moderately large ($r_m > .53$) and significant ($p < .05$) interaction between background noise and level of ability such that, as background noise increased, the performance of the pathetic students dropped more rapidly than the performance of the adequate students.

The treatments-by-levels design is not quite as simple as the independent-groups design, since it requires pretesting or matching of the subjects within levels, but it does provide information concerning the differences among the levels and the interaction between these levels and the independent variable. The procedure is very useful when there are large differences among the subjects with regard to a characteristic relevant to the variable being tested (the dependent variable). The problems of selecting the appropriate criteria for matching, which were discussed with regard to the Wilcoxon test, or the t test for matched groups, also apply to this design. The students used in this example, who differed so widely in mathematical ability, might be quite homogeneous for a simple independent-groups design if the measure were liver metabolism, reaction time, or learning of nonsense syllables.

There are other situations in which the treatments-by-levels design is particularly well suited, such as when subjects are obtained from populations that are basically different. Different populations might occur when testing students from two schools of unequal quality, from classes in different grades, or from different socioeconomic backgrounds or when testing mice of two different strains. In each of these cases the use of the treatments-by-levels design (with the samples from each presumed population constituting a level) would permit the examination of the effects of the major independent variable as well as the testing for differences among the levels and the interaction between these two factors. In this type of situation the interaction is often the most interesting effect. Therefore the treatments-by-levels design is particularly useful when the subjects are highly variable or when it is important to measure the interaction between various levels of a given characteristic and the main independent variable. For example, in studies of social behavior (of both human beings and animals), of personality factors, and of educational processes, we often are especially interested in the interaction between various levels of a given characteristic (e.g., group size, introversion, IQ) and the experimental manipulation.

FURTHER NOTES ON ANALYSIS OF VARIANCE

The restrictions on use of the analysis of variance are necessarily the same as those for the t test. In general, the data should be suitable for the use of means and standard deviations as measures of the scores within each group. Procedures for transforming skewed data to make it amenable to analysis of variance are briefly discussed in Supplement B. The formulas given in this chapter for two-way analysis of variance are suitable *only* when there is an equal number of observations within each cell. Unequal cell frequencies require a correction, which is cumbersome, but it can be most conveniently accomplished with one of the available computer programs. Suitable procedures for handling this correction are outlined in advanced texts (e.g., Winer 1962; Kirk 1968). Experiments that are designed to have an equal number of subjects in each cell may not end up that way if, for instance, subjects such as college sophomores fail to appear for the scheduled test or laboratory animals take sick and die.

The single-factor analysis of variance is not affected by unequal numbers of subjects in each group (just as the t test tolerates unequal sample sizes). The treatment-by-subjects design necessarily has an equal number of observations under each condition.

The analysis-of-variance procedures discussed in this and the previous chapter are the basic ones; there are many other designs that can be used in more complex situations. Such designs are, essentially, combinations of the independent-groups and repeated-measures designs discussed here and are understandable in the same terms.

EFFICIENCY OF THE FACTORIAL DESIGN

Among the advantages of the factorial design is that it is very economical of subjects. In the example of Figure 11.1 we compared the effects of different noise levels on the basis of three groups of six subjects each. We also compared the effects of easy and difficult problems with two groups of nine subjects each. And finally, we measured the interaction based upon the six cells of three subjects each. To do these studies with a simple analysis-of-variance design, it would have been necessary to use nearly twice as many subjects in two separate experiments just to measure the effects of noise level and task difficulty, and, even with the larger number of subjects, there would have been no measure of interaction. Therefore, the factorial design economically extracts multiple information from each subject by measuring both the major variables and then provides a measure of the interaction for little additional cost. The efficiency and popularity of the two-way analysis-of-variance design, therefore, lies in the large amount of information obtained from a relatively small number of subjects.

TWO-WAY ANALYSIS OF VARIANCE
(Factorial Design, Treatments-by-Levels Design)

Formulas

Source	df	SS	MS	F
Columns	$C - 1$	$\sum \left(\dfrac{T_{C_i}{}^2}{N_{C_i}} \right) - \dfrac{T^2}{N}$	$\dfrac{SS_C}{df_C}$	$\dfrac{MS_C}{MS_W}$
Rows	$R - 1$	$\sum \left(\dfrac{T_{R_j}{}^2}{N_{R_j}} \right) - \dfrac{T^2}{N}$	$\dfrac{SS_R}{df_R}$	$\dfrac{MS_R}{MS_W}$
(Cells)	—	$\sum \sum \left(\dfrac{T_{C_i R_j}^2}{N_{C_i R_j}} \right) - \dfrac{T^2}{N}$	—	—
$C \times R$ Interaction	$(C - 1)(R - 1)$	$SS_{\text{cells}} - SS_C - SS_R$	$\dfrac{SS_{CR}}{df_{CR}}$	$\dfrac{MS_{CR}}{MS_W}$
Within (Error)	$N - RC$	$SS_T - SS_{\text{cells}}$	$\dfrac{SS_W}{df_W}$	—
Total	$N - 1$	$\sum \sum (X^2) - \dfrac{T^2}{N}$	—	—

$$\text{where} \quad \frac{T^2}{N} = \frac{(\sum \sum X)^2}{N}$$

Table E   provides p values based on the F ratio, df for the numerator (df_N), and df for the denominator (df_D).

Table C   provides r_m values based on the F ratio, df_N, and df_D.

Restrictions   The scores within each cell should be close to normally distributed with the cell standard deviations being roughly equal. The effect of skewness, or unequal variance, decreases as the sample sizes increase.

Procedure   1. For each cell obtain the value of $\sum X \, (= T_{C_i R_j})$ and $\sum (X^2)$.
2. Calculate T^2/N by summing all the $\sum X$ terms from

step 1 to obtain the sum of all the scores. Square this sum and divide by the total number of scores (N).

3. Calculate the total SS term (SS_T) by adding the $\sum (X^2)$ terms obtained in step 1 and subtract T^2/N.

4. Calculate the columns SS term (SS_C) by adding the $T_{C_i R_j}$ terms for the cells in each column to obtain the sum of all the scores in each column. Square the sum for each column and divide by the number of scores in the column. Add the resulting terms for all columns and subtract T^2/N.

5. Calculate the rows SS term (SS_R) by adding the $T_{C_i R_j}$ terms for the cells in each row to obtain the sum of all the scores in each row. Square the sum for each row and divide by the number of scores in the row. Add the resulting terms for all rows and subtract T^2/N.

6. Calculate the cells SS term (SS_{cells}) by squaring the $T_{C_i R_j}$ term for each cell and dividing $T^2_{C_i R_j}$ by the number of scores in the cell. Add the resulting terms for all cells and subtract T^2/N.

7. The interaction SS (SS_{CR}) is obtained by subtracting both SS_R and SS_C from SS_{cells}.

8. The within, or error, term (SS_W) is obtained by subtracting SS_{cells} from SS_T.

9. Using the appropriate formula, calculate the df for each term.

10. Divide SS_C by df_C to obtain MS_C.
    Divide SS_R by df_R to obtain MS_R.
    Divide SS_{CR} by df_{CR} to obtain MS_{CR}.
    Divide SS_W by df_W to obtain MS_W.

11. Divide MS_C, MS_R, and MS_{CR}, respectively, by MS_W to obtain the F ratios for columns, rows, and interaction.

12. Use Table E to determine the p values for each F ratio based on F, df_N, and df_D.

13. Use Table C to determine the r_m value for each F ratio based on F, df_N, and df_D.

14. The interpretation of the analysis involves separate consideration of the rows effect, the columns effect, and the interaction.

Example

Subjects with low-, medium-, or high-drive levels are given either easy or difficult lists of nonsense syllables to learn. Trials to a criterion of one perfect trial is the performance measure

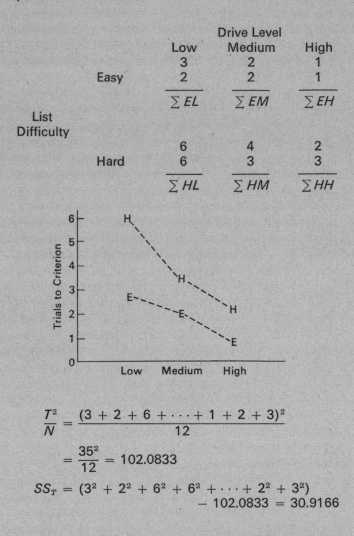

		Drive Level		
		Low	Medium	High
	Easy	3	2	1
		2	2	1
		$\sum EL$	$\sum EM$	$\sum EH$
List Difficulty				
	Hard	6	4	2
		6	3	3
		$\sum HL$	$\sum HM$	$\sum HH$

$$\frac{T^2}{N} = \frac{(3 + 2 + 6 + \cdots + 1 + 2 + 3)^2}{12}$$

$$= \frac{35^2}{12} = 102.0833$$

$$SS_T = (3^2 + 2^2 + 6^2 + 6^2 + \cdots + 2^2 + 3^2) - 102.0833 = 30.9166$$

$$SS_C = \frac{(3 + 2 + 6 + 6)^2}{4} + \frac{(2 + 2 + 4 + 3)^2}{4}$$
$$+ \cdots - 102.0833 = 12.6667$$
$$SS_R = \frac{(3 + 2 + 2 + 2 + 1 + 1)^2}{6}$$
$$+ \cdots - 102.0833 = 14.0834$$
$$SS_{cells} = \frac{(3 + 2)^2}{2} + \frac{(6 + 6)^2}{2} + \cdots + \frac{(2 + 3)^2}{2}$$
$$- 102.0833 = 29.4167$$

Source	df	SS	MS	F	p	r_m
Columns	2	12.6667	6.3333	25.3332	<.01	>.80
Rows	1	14.0834	14.0834	56.3336	<.001	>.80
(Cells)		29.4167				
$C \times R$ Interaction	2	2.6667	1.3333	5.3332	<.05	>.75
Within (Error)	6	1.5000	.2500			
Total	11	30.9166				

Conclusion

Columns: There are large ($r_m > .80$) and significant ($p < .01$) differences in trials to criterion among the three drive levels with the slowest learning occurring under the low-drive condition and the fastest under high drive.

Rows: The easy list tends to be learned significantly ($p < .001$) and appreciably ($r_m > .80$) more rapidly than the hard list.

Interaction: There is a large ($r_m > .75$) and significant ($p < .05$) interaction between drive level and list difficulty. Increasing drive level leads to a greater improvement in learning the hard list than in learning the easy list.

TREATMENTS-BY-SUBJECTS ANALYSIS OF VARIANCE

Formulas

Source	df	SS	MS	F
Columns	$C - 1$	$\sum \left(\dfrac{T_{C_i}{}^2}{N_{C_i}}\right) - \dfrac{T^2}{N}$	$\dfrac{SS_C}{df_C}$	$\dfrac{MS_C}{MS_e}$
Rows (Subjects)	$R - 1$	$\sum \left(\dfrac{T_{R_j}{}^2}{N_{R_j}}\right) - \dfrac{T^2}{N}$	—	—
Residual (Error)	$(C - 1)(R - 1)$	$SS_T - SS_C - SS_R$	$\dfrac{SS_e}{df_e}$	—
Total	$CR - 1$	$\sum \sum (X^2) - \dfrac{T^2}{N}$	—	—

$$\text{where } \frac{T^2}{N} = \frac{(\sum \sum X)^2}{N}$$

Table E   provides p values based on the F ratio, df for treatments (df_N), and df for the error term (df_D).

Table C   provides r_m values based on the F ratio, df_N, and df_D.

Restrictions   The scores within each condition should be close to normally distributed with the mean and standard deviation being suitable measures. The variance of the samples should be roughly equal. The effect of skewness or of unequal variance decreases as the sample sizes increase. When the scores are highly skewed and the sample size is not large, the Friedman test (based on ranks) may be preferable to the analysis-of-variance procedure.

Procedure
1. Calculate T^2/N by obtaining the sum of all the scores (T). Square this sum and divide by the total number of scores (N).
2. Calculate the total SS term (SS_T) by adding the square of each score to obtain $\sum (X^2)$ and subtract T^2/N.
3. Calculate the columns SS term (SS_C) by adding the scores in each column (i.e., adding across

subjects). Square the sum for each column ($T_{C_i}^2$) and divide by the number of scores in the column. Add the resulting terms for all columns and subtract T^2/N.

4. Calculate the subjects, or rows, SS term (SS_R) by adding all the scores for each subject. Square each of these sums ($T_{R_i}^2$) and add the squared values across subjects. Divide by N_{R_i}. Subtract T^2/N.
5. Calculate the residual error term SS (SS_e) by subtracting both SS_C and SS_R from SS_T.
6. Calculate df for each term according to the formula where C = number of columns, or conditions and R = number of rows, or subjects.
7. Obtain the MS for columns (MS_C) by dividing SS_C by df_C. Obtain the MS for the error term (MS_e) by dividing SS_e by df_e.
8. Divide MS_C by MS_e to obtain the F ratio.
9. Use Table E to determine p based on the F ratio, df_C (as df_N), and df_e (as df_D).
10. Use Table C to determine r_m based on the F ratio, df_N, and df_D.
11. If a test of subject differences is desired, follow steps 7 to 10 using SS_R and df_R in place of SS_C and df_C.

Example

To test for the effects of practice on the ability to solve simple problems, three two-minute math tests were administered to each subject. The problems on each test were of equivalent difficulty, and the second and third tests were preceded by five-minute rest periods. The measure was the number of problems solved in each test.

Subject	Test 1	Test 2	Test 3	R
a	5	7	10	22 $\sum R_a$
b	4	11	9	24 $\sum R_b$
c	8	5	12	25 $\sum R_c$
d	6	15	11	32 $\sum R_d$
e	9	12	16	37 $\sum R_e$
N = 5	$\sum C_1 = 32$	$\sum C_2 = 50$	$\sum C_3 = 58$	T = 140

$$\frac{T^2}{N} = \frac{(\sum R_a + \sum R_b + \cdots + \sum R_e)^2}{N}$$

$$= \frac{(5 + 7 + 10 + \cdots + 9 + 12 + 16)^2}{N}$$

$$= \frac{(140)^2}{15} = 1{,}306.6667$$

$$SS_T = (5^2 + 7^2 + 10^2 + \cdots + 12^2 + 11^2 + 16^2)$$

$$- \frac{T^2}{N} = 1{,}488 - 1{,}306.6667 = 181.3333$$

$$SS_C = \frac{(32)^2}{5} + \frac{(50)^2}{5} + \frac{(58)^2}{5} - \frac{T^2}{N}$$

$$= 1{,}377.6 - 1{,}306.6667 = 70.9333$$

$$SS_R = \frac{(22)^2}{3} + \frac{(24)^2}{3} + \frac{(25)^2}{3} + \frac{(32)^2}{3} + \frac{(37)^2}{3}$$

$$- 1{,}306.6667 = 1{,}359.3333 - 1{,}306.6667$$

$$= 52.6667$$

$$SS_e = SS_T - SS_C - SS_R = 181.3333 - 70.9333$$
$$- 52.6667 = 57.7333$$

Source	df	SS	MS	F	p	r_m
Columns	3 − 1 = 2	70.9333	35.4667	4.91	<.05	>.70
Rows (Subjects)	5 − 1 = 4	52.6667	—	—		
Residual (Error)	4 × 2 = 8	57.7333	7.2166			
Total	15 − 1 = 14	181.3333				

Conclusion   The number of problems solved differs significantly ($p < .05$) and appreciably ($r_m > .70$) across the three test sessions with the most problems being solved in the third session and the fewest in the first session.

SUMMARY

1. The *two-factor analysis of variance*, which is an extension of the single-factor analysis, handles data of two factors run simultaneously in one experiment.

2. The total variability is divided into portions attributable to each factor and to the error term. In addition, a portion of the variability can be attributed to the effect of one variable on the other variable—that is, to the *interaction* of the variables.

3. A related-samples test with more than two conditions has the form of the Friedman test and can be viewed as a two-factor design in which one of the factors (the rows) represents subjects. This design, an extension of the *t* test for related samples, is the *treatments-by-subjects design*. It does not permit measuring the interaction between the experimental variable and subjects.

4. The error term can be viewed as the variance not attributable to any of the experimental variables or to the interaction—that is, the variance that is "left over" or unaccounted for, the *residual variance*.

5. The *treatments-by-levels design* is a two-way analysis in which the rows condition represents homogeneous, or matched, groups of subjects. This procedure combines many of the advantages of both independent- and related-samples designs and permits measuring the interaction between the experimental variable and subject level.

chapter twelve

IMPORTANCE VERSUS PROBABILITY

Assume that you are considering a particular experiment but are not confident it is worth conducting. You seek advice from a colleague, explaining the experiment to him and then asking, "Do you think I ought to try it?" Now if the answer is merely Yes or No, you have not made much progress, since the basis of the judgment has not been given. A more effective procedure would be to question two groups of people, asking one group "How *likely* is it that the experiment will work?" and asking the other, "If the experiment works, how *important* will it be?" The answers will provide reasonable grounds for deciding whether or not to proceed with the experiment. With the original question, a No answer does not indicate whether the person thought the experiment was likely to work out but would be trivial or that it was important but very unlikely to succeed. When the question of likelihood, or probability, is separated from the question of importance, there is a much firmer basis for deciding how to proceed. Conceivably, in the absence of astute and informed confederates, you might profit from your own answers to these two questions. The final decision about the experiment involves an additional estimate of the cost of the project in terms of time, effort, and money. An indirect cost factor to keep in mind is that, when time, energy, money, or subjects are limited, conducting a given investigation delays or precludes another study.

Consider the following situation: One day you observe a physics professor in his laboratory; he is tossing an apple into the air, catching it, and tossing it up again—evidently with great interest and patience. You ask him what he is doing, and he replies (without pausing in his labor), "I think that Newton may have been wrong. If just one of these times the apple will stay up there,

197

I will go down in history." You readily concede the scientific importance of having the apple stay suspended in the air. However, you privately note that the probability of its happening is so remote that the experiment is essentially worthless (i.e., probability × importance $\cong$ 0). You heartily assure the good professor that he is well on the way to establishing his reputation and quietly leave the lab.

We can now turn to a rather dissimilar situation, that of the Anglo-Saxon jury trial previously discussed in Chapter 3. Typically, the jury is assigned the duty of determining only the defendant's guilt—that is, determining the probability that the accused is guilty. The judge has the problem, if the jury finds the defendant guilty, of assigning the verdict—that is, of assessing the importance of the crime and deciding upon an appropriate punishment. Therefore, there is an explicit separation between the assigning of probabilities to a particular event (done by the jury) and the attributing of importance to that event (done by the judge).

The question of assigning importance to scientific data has been discussed previously in terms of obtaining a measure of the size of the experimental effect (r_m) and probability values (p) for the data. The size of the effect is made explicit by the use of r_m, and the likelihood of the effect being due to chance is made explicit by the p value.

This chapter discusses how to obtain and apply the r_m statistic as an appropriate measure of the size of the differences among the groups. This measure, which is basically familiar from the material already covered, is also helpful in dealing with certain questions of experimental tactics and the choice of an appropriate sample size.*

†† MEASURES OF MAGNITUDE OF EFFECT

We have been dealing with problems of magnitude of experimental effect rather arbitrarily, using Table B with a measure of sample size and the probability value. We have also been referring to "high" and "low" levels of r_m without understanding the basis of these designations. To see how the r_m statistic is derived, we first have to review correlation.

Figure 12.1 is identical to Figure 8.5. Based on the formula for r^2, the correlation coefficient can be interpreted in terms of the ratio of the variance, or spread of the points about the best-fit line, compared to the variance, or spread, of the original scores. In this situation both sets of measurements are on a *continuous* scale, and any given score within the range can be obtained. A typical correlation study deals with such measurements as height and weight, and both of these variables can be considered to be *dependent* variables in the sense that neither was manipulated and both were measured.

* A portion of the following material is taken from H. Friedman, Magnitude of experimental effect and a table for its rapid estimation, *Psychological Bulletin*, 1968, **70** (4), 245–251, and reprinted by permission of the American Psychological Association; and H. Friedman, A simplified table for the estimation of magnitude of experimental effect, *Psychonomic Science*, 1969, **14** (4), 193.

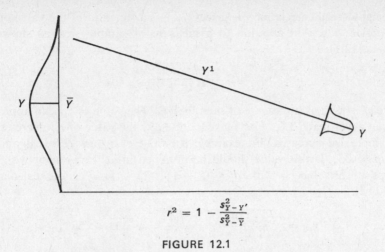

$$r^2 = 1 - \frac{s^2_{Y-Y'}}{s^2_{Y-\bar{Y}}}$$

FIGURE 12.1

The situation is slightly different when one of the variables is *dichotomized*, or divided into two discrete categories or levels. For example, we could have two groups of rats, one hungry and the other satiated (the dichotomized variable), and measure the speed with which each rat presses a bar in a Skinner box to obtain food (the continuous variable). In the case of height and weight we were concerned with predicting the individual's weight on the basis of his height (or vice versa). In the example with the rats, the independent, or experimental, variable—the degree of hunger—is manipulated in order to determine its effect upon the dependent variable (the rate of response). Now we want to know how well the rat's rate of response can be predicted if we know how hungry he is; or how well the rat's hunger can be determined, that is, to which group he belongs, if we know how fast he presses the bar.

A special version of the Pearson correlation coefficient deals with this type of problem. When the correlation coefficient is high, we can fairly accurately predict a rat's score in the Skinner box if we know which group he belongs to, or we can guess which group he belongs to if we know his score. High rates tend to be concentrated in the hungry group and low rates in the satiated group. On the other hand, if the correlation coefficient is low, we cannot accurately predict the rat's score on the basis of his group or guess his group on the basis of his score. In this case the difference between the scores of the two groups is relatively small compared to the spread of scores within each group; the groups, therefore, hardly differ at all.

††Point-Biserial Correlation for *t*

The data from the hungry-rat experiment can also be analyzed by a *t* test. The *t* test is similarly based on a measure of the difference between the groups (*DM*) compared to a measure of the spread of scores within each group (s_{DM}). It is not surprising, therefore, that the appropriate correlation measure, the **point-biserial correlation coefficient** (r_{pb}), is a function of the value of the appropriate t test in addition to a measure of sample size, as shown in formula (12.1):

$$r_{pb} = \sqrt{\frac{t^2}{t^2 + df}} \tag{12.1}$$

where df is used as a measure of sample size. The value of r_{pb} can approach but can never reach 1.0. For a given value of t the value of r_{pb} decreases as the sample size increases. For example, a t value of around 2.0 is significant with $p = .05$. If this value should be obtained in an experiment with two groups of 5 subjects, with $df = (5 - 1) + (5 - 1) = 8$, the calculation would be:

$$r_{pb} = \sqrt{\frac{(2.0)^2}{(2.0)^2 + 8}} = \sqrt{\frac{4}{12}} = \sqrt{.333} = .58 \tag{12.2}$$

indicating a moderately large effect.

Table 12.1 summarizes the change in r_{pb} when the t value remains constant and the sample size increases. The probability value for the t value is only slightly affected by the sample size (although enough so for the t distribution to be used in place of the normal distribution), whereas the r_{pb} values in the table clearly show that the size of the effect, or difference, is an *inverse* function of the number of subjects needed to detect that difference. In other words, large differences can be detected with relatively few subjects, whereas many subjects are required to detect small differences. (Table 2.3 illustrated the same point with χ^2.) The sample size selected for a study effectively sets a limit on the smallest effect that can be detected as being significant, and therefore, the sample size employed should be large enough for detection of the smallest difference that would be of interest. Later in the chapter the selection of sample sizes will be discussed in more detail.

We can also interpret r_{pb} in terms of a scattergram, but all the scores will necessarily fall at one of two points along the best-fit line rather than all along the line when two variables are continuous. The appropriate scatter-

TABLE 12.1. r_{pb} Values As a Function of df When $t = 2.0$

df	t	r_{pb}
6	2.0	.63
8	2.0	.58
10	2.0	.53
20	2.0	.41
30	2.0	.34
50	2.0	.27
100	2.0	.20
200	2.0	.14
500	2.0	.09

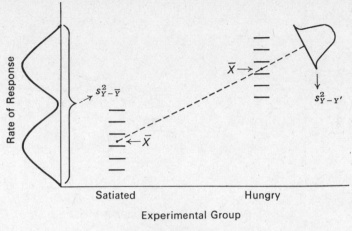

Experimental Group

FIGURE 12.2

gram for the hungry-rat example is shown in Figure 12.2. As indicated by the scattergram, the value of r_{pb} is measured in terms of $s^2_{Y-Y'}$ or the spread of the points about the means of each of the two groups (rather than in terms of the spread of points about the entire length of the best-fit line), relative to the total spread of scores, or $s^2_{Y-\bar{Y}}$. Taken as a correlation coefficient, the significance of the obtained r_{pb} is determined by the probability value of the associated t value. Therefore, a r_{pb} value is significant if it is based on a significant t value.

The r_{pb} value is appropriate as a measure of the difference between the scores of the two groups, or the size of the experimental effect. Since r_{pb} is a correlation measure, the improvement in our ability to predict a rat's score on the basis of knowing his group, in terms of the reduction of $s^2_{Y-Y'}$ relative to $s^2_{Y-\bar{Y}}$, is given by r_{pb}^2. For example, $r_{pb} = .50$ represents a moderate improvement in our ability to predict the rat's score ($r_{pb}^2 = .50^2 = .25$) and $r_{pb} = .30$ a small improvement ($r_{pb}^2 = .09$). In slightly different terms, r_{pb} can be viewed as a measure of the degree to which the independent variable affects the dependent variable. For example, if the response rate were not affected by drive level, then a knowledge of how hungry a rat is would not help in predicting his rate of response. The scores of the hungry and satiated rats would be virtually the same and all the values of t and r_{pb} would be close to 0.

††Eta for Analysis of Variance

The analysis of variance is similar to the t test, except that there may be more than two conditions, providing a scattergram like that shown in Figure 12.3. Here, too, we are concerned with the spread of scores about the means of each group compared to the total spread of scores. The appropriate statistic is based on the value of the F ratio; the df associated with the numerator, or numbers of groups (df_N); and the sample size in terms of the df

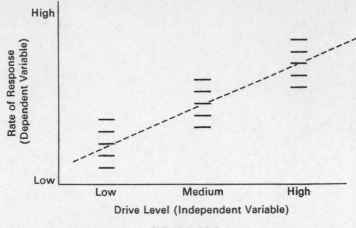

FIGURE 12.3

associated with the denominator (df_D). This measure, called **eta** (r_η) is defined as follows:

$$r_\eta = \sqrt{\frac{df_N F}{df_N F + df_D}} \qquad (12.3)$$

When there are only two groups, df_N is equal to 1, and since with two groups $F = t^2$, the formula for r_η is identical to that for r_{pb}.*

††Contingency Coefficient for χ^2

For categorical data, which are appropriately analyzed by χ^2, a measure of the magnitude of effect is given by the **contingency coefficient** (C), which is defined in terms of χ^2 and the sample size (N):†

$$C = \sqrt{\frac{\chi^2}{\chi^2 + N}} \qquad (12.4)$$

Formula (12.4) clearly has the same general form as r_η and r_{pb}, even though it is not directly related to these measures.

The statistics discussed in this section can be considered as comparable measures of the magnitude of the observed effect, and a single term (r_m) can be used, defined in terms of an inferential statistic (Q) and a measure of sample size (S):

$$r_m = \sqrt{\frac{Q^2}{Q^2 + S}} \qquad (12.5)$$

---

* Formula (12.3) can lead to an overestimation of the size of effect when applied to certain multi-factor or high-order analysis-of-variance designs. Fleiss (1969) provides alternative measures for these situations.

† The contingency coefficient is discussed further in a note at the end of the chapter.

Table C and r_m 203

where the appropriate measure for S, or sample size, for each statistic is given in Table 12.2.

TABLE 12.2

Statistic		Sample Size
Q	Q^2	S
t		df
Z		N
	$df_N(F)$	df_D
	χ^2	N

TABLE C AND r_m

Table C gives values of the statistic (Q) for levels of r_m from .20 through .80 and for sample sizes (S) from 5 through 200. Since the probability values of the t statistic are a function of df, the closest t values in the table equivalent to $p = .05$ and $p = .01$ are shown with an asterisk and a dagger, respectively. The marked values are not sufficiently precise to be used in place of the values given in Table A.

With a t test the r_m value is a function of the difference between the means (DM) in relation to the standard deviation $(\sigma$ or $S)$, as shown in the discussion of r_{pb}. When the two samples have equal size and equal variance, r_m values can be expressed directly as DM/σ. These values are given in Table C for each level of r_m. When the samples have unequal size or variance, the (DM/σ) values are slightly different.

To return to the analysis of variance, it can be shown that:

$$r_m = r_\eta = \sqrt{\frac{df_N(F)}{df_N(F) + df_D}} = \sqrt{\frac{SS_C}{SS_C + SS_W}} \tag{12.6}$$

Since in single-factor analysis of variance:

$$SS_C + SS_W = SS_T \tag{12.7}$$

$$r_m = \sqrt{\frac{SS_C}{SS_T}} \tag{12.8}$$

For the single-factor design, then, r_m reflects the proportion of the total variability (SS_T) attributable to the difference among groups—that is, to the effect of the independent variable (SS_C). Since the single-factor analysis of variance for two groups is identical to the t test, r_{pb} can also be understood in these same terms.

Another way of considering r_m is in terms of misclassification. When a difference between the scores of two groups is found, we can ask the question, "How well can the subjects be divided on the basis of the scores?" For

TABLE 12.3

$r_m = .80$

$r_m =$	.75		.8	
S	Q^2	Q	Q^2	Q
1.	1.29	1.13	1.78	1.33
2.	2.57	1.60	3.56	1.89
3.	3.86	1.96	5.33	2.31
4.	5.14	2.27	7.11	2.67
5.	6.43	2.54	8.89	2.98*
6.	7.71	2.78*	10.67	3.27
7.	9.00	3.00	12.44	3.53 †
8.	10.29	3.21	14.22	3.77
9.	11.57	3.40†	16.00	4.00
10.	12.86	3.59	17.78	4.22

$df = 8$ Closest table value
                      to $t = 3.73$

Dagger indicates that entries
below this point in the
column have $p < .01$.

example, if we know the speed with which each rat presses the bar, how accurately can we divide the rats into the hungry and the satiated groups? Misclassification will occur when an individual's score deviates from the mean of his group more than halfway toward the mean of the other group. In other words, if the total group is divided on the basis of the scores, such an individual will be included in the wrong group. With groups having equal size and variance, the proportion of subjects misclassified, $P(m)$, is a function of r_m, and the value of $P(m)$ for each level of r_m is shown at the bottom of Table C. The $P(m)$ value also applies to the main effects of analysis of variance when the groups are of equal size and df_D is large compared to df_N.

To summarize the use of Table C, let us return to the example of the children putting toys away and the data of Table 9.2, which are analyzed in equations (9.6) to (9.8). We found that for $t = 3.73$ and $df = 8$, $p < .01$ and $r_m > .76$ (taken from Table B). We now enter $S = df = 8$ and $Q = t = 3.73$ in Table C, the appropriate portion of which is shown in Table 12.3. The entry of 3.77 is sufficiently close and gives a value of $r_m = .8$, which agrees with the value from Table B. This entry is below the column entry marked with the dagger indicating significance at the .01 level. In addition, an r_m value equal to .8 is associated with $DM/\sigma = 2.67$ (shown at the bottom of Table C). The actual data for the two groups (with the variances averaged, since they are not equal) indicate that the difference in means divided by the calculated standard deviation is equal to 2.36, which is reasonably close to the table value. Table C further indicates that, if we should guess whether each child is in the control or the experimental group on the basis of the number of toys he put away, we would be in error approximately 9 percent of the time. In this example, then, the difference between the groups is rather large—as it must be to be detectable with as few as 10 subjects.

Table C can also be used to give approximate r_m values for nonparametric tests when these tests can be approximated by the χ^2 distribution (as in the Kruskal-Wallis and Friedman tests) or by the normal distribution (as in the binomial and White tests). When large sample approximations are not used, Table B provides usable r_m values based on the sample size and the probability values.

Whether or not a given value of r_m is interpreted as representing an important effect is necessarily a function of the particular situation. If r_m is being calculated as a measure of the degree to which the independent variable is a factor in the behavior being observed, then in situations in which other relevant variables are largely eliminated or held constant, as in a typical animal-learning experiment, rather large values of r_m can be expected. However, in situations involving numerous important variables, such as tests of academic performance as a function of intelligence, motivation, social pressure, and school quality, no single variable will be a highly effective basis for prediction as is drive or reward level when rats are being tested in a straight runway. As with the probability value, the value of r_m cannot suitably be discussed without reference to the context in which the experiment or study is being carried out.

SAMPLE SIZE AND EXPERIMENTAL TACTICS

One use of r_m is as a basis for comparing the outcomes of studies with different sample sizes, since the probability values alone are not a sufficient guide. Consider a situation in which the means of the control and experimental groups differ by $1.01\ \sigma$ (i.e., $DM/\sigma = 1.01$). The column in Table C for $DM/\sigma = 1.01$ shows that $r_m = .45$ and, with $df = 18$, $t = 2.14$ with $p < .05$. If the sample size is increased to 38 ($df = 36$) with DM/σ unchanged, then $t = 3.02$ and $p < .01$, but r_m is still .45. The t value increases with greater sample size (specifically, t^2 increases as a linear function of df for a given r_m value), and the probability value decreases accordingly. However, the value for r_m, which takes sample size into account, remains constant and reflects the same magnitude of effect observed in both cases. Therefore, when two studies are compared, the one demonstrating the larger effect is the one with the larger r_m value, not necessarily the study with the lower (more significant) p value.

The r_m table can also be used to find the sample size needed to detect a given effect. For example, if the expected difference between group means is 1 standard deviation ($DM/\sigma = 1.0$), then (with a t test) we are dealing with a relationship on the order of $r_m = .45$. The column for $r_m = .45$ indicates that this relationship can be expected to give a significant t value at the .05 level with $df = 18$, or two sample sizes of 10 subjects each. Of the other hand, if the difference expected is small, less than one-half of the standard deviation or $r_m = .2$, a sample size of approximately 100 is needed to reject the null hypothesis at the .05 level. But how do we know what size effect, or difference, to expect? Data from a pilot, or preliminary, study provide a reasonable estimate of the r_m value to be expected in subsequent experiments. Similarly, if a planned investigation is a near replication of a previous study, the prior r_m value will be expected. Frequently, published reports of studies similar to a planned experiment will allow estimating the expected r_m value.

The r_m table also provides a very convenient way of dealing with a related and important aspect of experimental tactics—deciding what to do after the initial data of a study have been collected. Consider a situation in which analysis of the first 12 subjects in each of two groups gives a t value for the difference between means of 1.48, which for $df = 22$ yields $r_m = .3$. We could *assume* that this value is an accurate measure of r_m, in which case a sample size of approximately 50 would be required in order for us to reject the null hypothesis at the .05 level, and an additional 26 or 30 subjects would be needed. However, if these data are rates of response of hungry and satiated groups of pigeons in a Skinner box, the value for r_m is quite low in view of the degree of control that is typical in this type of experiment. Therefore, although an increase in sample size might permit us to say that we had a significant increase in rate of response with an increase in drive level, we still would not be able to predict adequately the rate of response from a knowledge of drive level. On the basis of the information obtained from the first 24 subjects in this case, we could assume that the magnitude of effect in the study is too small to be acceptable. Perhaps we should modify the experiment in order to increase the effectiveness of the independent variable (i.e., increase r_m) rather than add additional subjects to the same experiment. With r_m increased to .5, 8 or 10 subjects in a group will probably be sufficient to obtain a significant effect. The primary ways to increase the value of r_m are to modify the manipulation of the independent variable (the task may be too difficult or too easy for the subjects or the range of stimuli may be too limited), to reduce the variability by controlling the action of extraneous variables (e.g., keeping the testing room quiet), or by more careful selection of subjects (e.g., using subjects with similar IQ, age, or background).

The results from a given experiment often present the alternatives of using more subjects to obtain a significant effect at a given r_m level or attempting to increase the value of r_m by modifying the study. As in the previous example an improved experiment may yield a significant effect with fewer subjects than would be needed if subjects were added to the original experiment. Since subjects can be added only if they are comparable to the original sample (and often this is not possible, as when fixed groups such as class sections are used), a replication using larger groups may be necessary. If the original data are assumed to give an accurate reflection of the effectiveness of the experimental manipulation, a suitable decision regarding the two alternatives can be made on the basis of the cost of additional subjects (rats and sophomores are generally cheap to use), the difficulty of improving the experiment, and the acceptable level of r_m. The main advantage in using the r_m table in dealing with this problem is that it makes the two alternatives readily comparable.

†† NOTE ON THE CONTINGENCY COEFFICIENT AND THE PHI COEFFICIENT

One disadvantage of the contingency coefficient as a measure of magnitude of effect is that the maximum value of C is a function of the df of the associated χ^2. When $df = 1$, the maximum possible value is $\sqrt{.5}$ or .71. As df increases, so does the maximum value for C. The result of this fact is that the comparison of r_m or C values for situations involving different df levels becomes awkward, particularly if the C values are large and approach the possible maximum. Low C values can safely be interpreted as being low, regardless of df level. Since the maximum is most limited when $df = 1$, the **phi coefficient** (r_ϕ), defined as

$$\text{phi} = r_\phi = \sqrt{\frac{\chi^2}{N}} \qquad \text{when } df = 1 \qquad (12.9)$$

is often used in preference to C for the 2×2 table. The maximum value of r_ϕ is 1.0, and this statistic is directly related to the Pearson correlation coefficient. In practice (particularly when N is large compared to χ^2), the values of r_ϕ and C tend to be fairly close. For example, when $\chi^2 = 5$ and $N = 25$, $r_\phi = .45$ and $C = .41$.

When calculating C or r_ϕ on the basis of χ^2 for $df = 1$, it is more appropriate to use χ^2 rather than χ_c^2. The correction for continuity used for χ_c^2 is specifically designed to adjust the probability values when N is small. The degree of relationship, r_m, is a direct function of χ^2 and not of χ_c^2. Therefore, the use of probability values based on χ_c^2 to obtain r_m values from Table B involves an approximation, but the effect upon r_m is generally small and decreases as N increases.

CONTINGENCY AND PHI COEFFICIENTS

Formulas

$$C = \sqrt{\frac{\chi^2}{\chi^2 + N}}$$

can be used when $df = 1$ or > 1.

$$r_\phi = \sqrt{\frac{\chi^2}{N}}$$

can be used only when $df = 1$ but is generally preferable to C. Based on χ^2 rather than χ_c^2.

Table F

provides p values based on χ^2 and df. The C or r_ϕ coefficients are significant if the associated χ^2 value is significant.

Table C

provides estimates of C (as r_m) based on χ^2 and N.

Restrictions

A disadvantage of the contingency coefficient is that the maximum value possible is less than 1.00. The value is most limited when $df = 1$. Comparison of C coefficients based on different df levels is imprecise, particularly when the C values are near the maximum.

The phi coefficient (r_ϕ) can only be used when $df = 1$. It has a range from .00 to 1.00 when based on equal-sized samples. When N is large relative to χ^2, the values of C and r_ϕ tend to be quite similar.

Example 1

C for $df > 1$
The majors of college men and women of equivalent ability, in terms of IQ and College Board scores, were studied, with the following results :

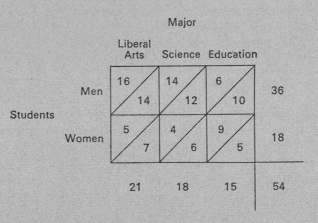

Given that $\chi^2 = 6.65$ with $N = 54$, $df = 2$.
Since $df = 2$, C must be used, not r_ϕ.

$$C = \sqrt{\frac{6.65}{6.65 + 54}} = \sqrt{\frac{6.65}{60.65}} = \sqrt{.1096} = .33$$

From Table C, where $r_m = C$, $.33 > C > .30$ based on $\chi^2 = 6.65$ with $N = 54$.
From Table F, $p < .05$ with $\chi^2 = 6.65$ and $df = 2$.

Conclusion There is a moderate ($C > .30$) and significant ($p < .05$) relationship between the students' sex and their major area of study, with men tending to major in liberal arts and science and women tending to prefer education.

Example 2 r_ϕ and C for $df = 1$
One statistics course is taught with teaching machines, while an equivalent course is taught with a standard text and lectures. The number of students that pass and fail a common final examination is measured.

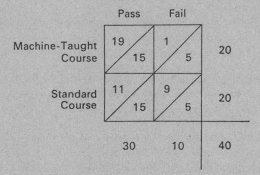

Given that $\chi^2 = 8.53$ with $N = 40$, $df = 1$.
Since $df = 1$, either r_ϕ or C can be used, r_ϕ being generally preferable.

$$r_\phi = \sqrt{\frac{8.53}{40}} = \sqrt{.2133} = .46$$

$$C = \sqrt{\frac{8.53}{8.53 + 40}} = \sqrt{.1758} = .42$$

From Table C, where $r_m - C$, $.45 > C > .40$ based on $\chi^2 = 8.53$ with $N = 40$.
From Table F, $p < .05$ based on $\chi^2 = 6.53$ with $df = 1$.
Note: The values of r_ϕ and C are similar when the χ_C^2 value is small relative to N.

Conclusion There is a significant ($p < .05$) and moderate relationship ($r_\phi = .46$ or $C = .42$) between the type of instruction and the number of students that pass or fail the course. The machine-taught section had a higher proportion of passing students.

209

SUMMARY

1. The interpretation of data from a study is aided by a measure of the size, or *importance*, of the effect (r_m) as well as a measure of the likelihood, or *probability* (p), of the data being a sample from the population described by H_o.

†† The r_m value for a t test is given by the *point-biserial correlation* (r_{pb}) coefficient, which is a measure of how well a subject's score can be predicted from the test group he belongs to.

†† With analysis of variance the appropriate measure of r_m is *eta* (r_n), which is an extension of the point-biserial coefficient for more than two groups.

†† The r_m value for χ^2 tests is based on the *contingency coefficient* (C), which has the same general form as r_{pb} and r_n.

2. Table C permits the rapid estimation of r_m from a measure of sample size and the value of the t, F, or χ^2 statistic. For each level of r_m, the table gives values based on the t test for the difference between group means in SD units and a measure of misclassification.

3. The *sample size* needed for a particular study, based on an estimate of the size of the effect to be obtained, can be derived from Table C. The initial data from a study often involve problems of *experimental tactics*, making it necessary to decide whether to add a sufficient number of subjects to obtain a significant effect or revise the experiment to make it more efficient. Table C provides a clear picture of the cost and value of these two alternatives.

†† A disadvantage of the contingency coefficient is that the maximum possible value is a function of *df*, or the number of cells. Caution has to be exercised when comparing C values (particularly high values) for χ^2 tests based on different *df*'s. When $df = 1$ (i.e., with a 2×2 array), the *phi coefficient* (r_ϕ), which has a maximum value $= 1$, may be preferable to C. In practice the values of r_ϕ and C are similar when N is large compared to the value of χ^2.

chapter thirteen

SUMMARY OF CRITERIA FOR SELECTING APPROPRIATE TEST

In the preceding chapters a variety of measures and tests that aid in the interpretation of research data was considered. When we design an experiment or study, the choice of the statistical analysis affects the selection of subjects as well as the procedure. If we delay thinking about the analysis until the data have been collected, we may find that the most informative test or measure cannot be used and that a small modification of the design would have made it possible. By way of summary, the questions asked of the data and the factors that enter into the choice of the appropriate statistical test are reviewed below.

The first step is to classify the experimental design: Is it a single-sample or multi-sample design?

The second step is to determine whether independent groups or related samples are to be employed. The choice between these two approaches is a function of the relative advantages and disadvantages discussed in Chapters 4, 9, and 11. When related samples are used, are we primarily interested in testing for differences in performance under different conditions or in the correlation of the different measures?

The third step is to decide if the data are to be in the form of category measurements or scores. If scores are to be used, can they be used directly, as interval data, or should they be ranked first, as ordinal data? The final decision about treating the data as scores or ranks cannot be made until after the data have been collected. With an analysis of variance the data must be treated directly as scores. Fortunately, as sample sizes increase and ranking tests become markedly more difficult to calculate, the interval-data tests become increasingly tolerant of skewness and unequal variability in the samples. (The choice between treating the data as interval or ordinal is discussed in the text and also in Supplement B on transformations.)

GUIDE TO OUTLINE ON INSIDE COVERS

The following outline lists these steps and shows the appropriate test for each situation considered in the text. This outline is given in a concise form on the inside covers where the pages containing a worked example and the pages of the necessary tables are given for each test mentioned.

I. Single-sample and goodness-of-fit tests
   Expected values are derived from prior knowledge or a theory.
   Categorical data: χ^2 or binomial tests
   Score data: t test
II. Two-sample tests
   A. Independent samples, or two separate groups
      Categorical data: χ^2 multi-sample test
      Ranked data: White test
      Score data: t test
   B. Related samples. Each subject measured twice or matched pairs of subjects employed
      1. Primary interest: the differences between scores of the two samples
         Ranked data: Wilcoxon signed-ranks test
         Score data: t test for related samples
      2. Primary interest: the correlation of scores between the two samples
         Ranked data: Rank-order correlation
         Score data: Pearson correlation and slope of the best-fit line
III. Three or more samples
   A. Independent samples
      Categorical data: χ^2 multi-sample test
      Ranked data: Kruskal-Wallis test
      Score data: single-factor analysis of variance
   B. Related samples
      Ranked data: Friedman test
      Score data: treatments-by-subjects analysis of variance
IV. Two-variables: factorial design or treatments-by-levels design
   Score data: two-way analysis of variance

Finally, we must remember that the calculation of p and r_m values is only an initial step in the analysis of data. These values aid in assessing the significance and importance of observed effects (differences or relationships), but the final interpretation depends upon a knowledge of the situation under investigation.

supplements

brief review of basic mathematics

Multiplication, Division, Addition, and Subtraction

1. Multiplication and division are carried out before addition and sub-traction (unless explicit instructions are given as in instruction 3).

   EXAMPLE: $4 \times 3 + 2 = 12 + 2 = 14$
   $3 + 9 \div 3 = 3 + 3 = 6$

2. When both division and multiplication or two or more division in-structions are involved, explicit instructions concerning grouping are indicated by enclosures or fraction bars (see instructions 3 and 4) to avoid ambiguity.

3. Parentheses ( ) or brackets [ ] indicate that the terms within the enclosure are to be treated as single values.

   EXAMPLE: $5(6 + 3) = 5(9) = 45$
   or $5(6) + 5(3) = 30 + 15 = 45$

   $$\frac{1}{2}(6 + 4) = \frac{1}{2}(10) = 5$$

   or $\frac{1}{2}(6) + \frac{1}{2}(4) = 3 + 2 - 5$

   Note that the multiplication or division applies to all the terms within the enclosure. Specifically, $A(B + C) = AB + AC$

4. A fraction bar (———) serves the same purpose as an enclosure. Terms above and below the bar are treated as single values.

   EXAMPLE: $\dfrac{6 + 4}{2} = \dfrac{10}{2} = 5 = \dfrac{1}{2}(6 + 4)$

   $\dfrac{6 + 4}{3 + 2} = \dfrac{10}{5} = 2 = \dfrac{1}{5}(6 + 4)$

5. A radical, or square root, sign serves as an enclosure. Addition and subtraction of terms under the radical sign must be carried out before extracting the root. (See the section on square roots for rules concerning multiplication and division.)

EXAMPLE: $\sqrt{16 + 9} = \sqrt{25} = 5$ $NOT \sqrt{16} + \sqrt{9}$

Fractions

1. Simple fractions. Fractions are unchanged in value when *both* the numerator (top term) and denominator (bottom term) are multiplied or divided by the same value (other than 0). This provides a convenient means of simplifying fractions.

EXAMPLE: $\dfrac{A}{B} = \dfrac{A/X}{B/X} = \dfrac{A(Y)}{B(Y)}$

$\dfrac{13}{104} = \dfrac{13/13}{104/13} = \dfrac{1}{8}$

$\dfrac{.03}{.11} = \dfrac{.03 \times 100}{.11 \times 100} = \dfrac{3}{11}$

2. Addition and subtraction. Fractions can be added and subtracted only if they have the same denominator, that is, a common denominator. The easiest way of obtaining a common denominator is to use the product of the denominators of the individual (simplified) fractions.

EXAMPLE: $\dfrac{a}{b} + \dfrac{c}{d} = \dfrac{ad}{bd} + \dfrac{bc}{bd} = \dfrac{ad + bc}{bd}$

or $\dfrac{3}{5} + \dfrac{2}{7} = \dfrac{3(7)}{5(7)} + \dfrac{(5)2}{(5)7} = \dfrac{21 + 10}{35} = \dfrac{31}{35}$.

or $\dfrac{2}{3} - \dfrac{2}{7} = \dfrac{2(7)}{3(7)} - \dfrac{(3)2}{(3)7} = \dfrac{14 - 6}{21} = \dfrac{8}{21}$

3. Multiplication. Fractions are multiplied by multiplying the numerators, which provides the numerator of the products, and multiplying the denominators, which provides the denominator of the product.

EXAMPLE: $\dfrac{A}{B} \times \dfrac{C}{D} = \dfrac{AC}{BD}$

or $\dfrac{5}{7} \times \dfrac{3}{4} = \dfrac{5(3)}{7(4)} = \dfrac{15}{28}$

or $\dfrac{5}{7} \times 9 = \dfrac{5}{7} \times \dfrac{9}{1} = \dfrac{45}{7} = 6\dfrac{3}{7}$

4. Division. To divide one fraction by another fraction, multiply both the numerator and the denominator by a common denominator to obtain a simple fraction.

EXAMPLE: $\dfrac{A/B}{C/D} = \dfrac{A/B(BD)}{C/D(BD)} = \dfrac{AD}{CB}$

Or, more simply, invert the denominator term and multiply by the numerator.

EXAMPLE: $\dfrac{A/B}{C/D} = \dfrac{A}{B} \times \dfrac{D}{C} = \dfrac{AD}{BC}$

$\dfrac{2/3}{4/5} = \dfrac{2}{3} \times \dfrac{5}{4} = \dfrac{10}{12}$

Decimal Points

1. Multiplication. The product of two or more terms contains as many decimal places as the *sum* of the decimal places of the individual terms.

EXAMPLE: $3.7 \times .54 = \dfrac{37}{10} \times \dfrac{54}{100} = \dfrac{1,998}{1,000} = 1.998$

2. Division. The decimal point of the divisor is moved to the right of the last nonzero digit after the decimal, and the decimal point of the dividend is moved the same number of places to the right. Standard long division is carried out and the decimal point of the quotient, or result, is placed in line with that of the dividend.

EXAMPLE: $\dfrac{205}{2.5} = \dfrac{2050.}{25.} = 25.\overline{)2050.}$

$$\begin{array}{r} 82. \\ 25.)\overline{2050.} \\ \underline{200} \\ 50 \\ \underline{50} \end{array}$$

Square Roots

When the square root ($\sqrt{\ }$) of a given number, X (i.e., $\sqrt{X}$), is multiplied by itself, the result equals the given number [i.e., $(\sqrt{X})(\sqrt{X}) = X$].

Square roots can be obtained by hand calculation (slowly) or automatically by some electronic calculators (instantly). The use of tables for obtaining square roots directly and the use of standard calculators for readily determining square roots will be covered in the next section.

Note that although $\sqrt{4} = 2$, $\sqrt{40} \neq 20$. Clearly, $\sqrt{400} = 20$ and, in fact, $\sqrt{40} = 6.32$. When the square root is multiplied by 10, the square is multiplied by 100. There are two types of numbers: those having an odd number of digits to the left of the decimal (e.g., $\sqrt{4} = 2$, $\sqrt{400} = 20$, $\sqrt{40,000} = 200$) and those with an even number of digits to the left of the decimal (e.g., $\sqrt{40} = 6.32$, $\sqrt{4,000} = 63.2$). We will deal with numbers smaller than 1 in a fashion similar to that for larger numbers.

To determine how many digits will be to the left of the decimal point in the square root of a number, mark off the number in pairs of digits, working in *both* directions from the decimal point.

EXAMPLE: $427.35 = 4\ 27.35$
$42.735 = 42.73\ 5$
$.42735 = .42\ 73\ 5$

We are specifically concerned with the left-most markings. These indicate whether we are dealing with a number of the even or odd group.

EXAMPLE:

Even		Odd	
X	$\sqrt{X}$	X	$\sqrt{X}$
12 34	35.13	1 23 .4	11.11
12 .34	3.513	1 .23 4	1.111
.12 34	.3513	.01 23 4	.1111
.00 12 34	.03513	.00 01 23 4	.0111

Once the digit pairs are marked off, the first digit of the square root should be determined to catch errors. For the even-digit group in the example, the first digit is based on $\sqrt{12}$ (the left-most pair), and 3 (for $\sqrt{9}$) is the closest value. Similarly, for the odd-digit set, the base is $\sqrt{1}$ or 1 as the first digit. The square root of a number larger than 1 will have as many digits to the left of the decimal point as there are *pairs* of digits in the number to the left of the decimal point. The square root of a number smaller than 1 will have as many zeros to the right of the decimal point as there are *pairs* of zeros in the number to the right of the decimal point.

Square-Root Tables
If the material in the preceding section is kept in mind, the table of square roots (Table O) will present no problems. Table O provides the square roots for all numbers from 1 to 1,000. The column headed $\sqrt{N}$ gives the value for the indicated number. Note that the square roots of numbers smaller than 1.0 are *larger* than the numbers (e.g., $\sqrt{.49} = .7$).

If the number has more than three digits, the table provides only a good estimate of the square root.

If greater accuracy is desired and a standard electric or electronic calculator is available, the Monroe method (Table Q) will usually be satisfactory: This method gives the square root accurately to five digits and with a small error, at most, in the sixth digit. The procedure is very simple:

1. Locate the number in column N' (keeping track of whether it is an odd- or even-digit number).
2. Add the associated number from column A to the number, placing the decimal point so the two numbers (column A and column N') are almost equal.
3. Enter the sum as a dividend at the extreme left of the keyboard or use the maximum decimal setting. (On certain machines the first value can be entered as the dividend and the second value added directly to it.)
4. Divide by the value from column D, setting the decimal point to make the value approximately *twice* the estimated square root.

TABLE A.1. Section of Table Q

N'	A	D	N'	A	D
1.045	1067	2066	6.49	6579	513
1.09	1114	2111	6.69	6791	5212
1.14	1162	2156	6.905	7017	5298
1.19	1211	2201	7.132	7244	5383
1.24	126	2245	7.35	745	5459
1.28	1303	2283	7.56	7681	5543
1.327	1356	2329	7.80	7924	563
1.385	1416	238	8.05	8168	5716
1.45	1481	2434	8.285	8404	5798
1.515	1545	2486	8.53	8661	5886
1.575	1609	2537	8.795	8922	5974
1.645	1677	259	9.00	9102	6034
1.705	1741	2639	9.24	9345	6114
1.777	1817	2696	9.485	9619	6203
1.855	1896	2754	9.77	9894	6291
1.94	1974	281	10.0	1009	6353
2.01	2042	2858	10.2	1032	6425
2.08	2114	2908	10.45	1053	649
2.16	2177	2951	10.6	1066	653
2.222	2259	3006	10.8	1086	6591
2.29	2327	3051	11.0	11122	667
2.38	2421	3112	11.2	1136	6741
2.47	252	3175	11.5	1157	6803
2.57	2621	3238	11.7	1183	6879
2.676	2729	3304	11.9	1202	6934
2.78	2829	3364	12.1	1218	698
2.885	2936	3427	12.3	1245	7057
2.99	3038	3486	12.5	1262	7105
3.09	3154	3552	12.75	1287	7175
3.22	3285	3625	12.9	1305	7225
3.352	3404	369	13.2	13286	729
3.457	3523	3754	13.4	1352	7354

Not here! 12.34 is an even group number

Odd Group

Even Group

Add 12.45

Divide by 7.057

12.34 is between these numbers

5. The quotient equals the square root after correct placement of the decimal point (the decimal point should be correct but it must be checked).

For example, $\sqrt{12.34}$ can be found using Table A.1 as follows:

1. The number is between 12.3 and 12.5 in column N'.
2. Column A = 12.45, added to 12.34 = 24.79.
3. Enter 24.79 as a dividend.
4. Divide by column D value (7.057). Note that 7.0 is about twice the estimated root.

5. Quotient = 3.5128. After the decimal point is checked, square root = 3.5128.

Logarithms

Common logarithms are the power to which the number 10 must be raised to equal a given number.

EXAMPLE: $10^2 = 100$; log 100 = 2
$10^3 = 1,000$; log 1,000 = 3

and

log 200 is greater than 2 and less than 3
log 200 = 2.3010

Since $10^1 \times 10^3 = 10^{1+3} = 10^4$ or $10 \times 1,000 = 10,000$, multiplication can be carried out by *adding* logarithms.

Similarly, division is reduced to *subtracting* logs.

EXAMPLE: $10,000/10 = 10^4 \div 10^1 = 10^3 = 1,000$

Since $(10^2)^3 = 10^{2 \times 3} = 10^6$, raising numbers to a given power can be accomplished by *multiplying* logs.

Therefore, extracting roots becomes a process of *dividing* logs.

EXAMPLE: $\sqrt{10,000} = \sqrt[2]{10^4} = \dfrac{10^4}{10^2} = 10^2 = 100$

The number for which a log is determined is the antilog. If the logarithms have been obtained and manipulated, the *antilog* can be obtained by reversing the process for obtaining the log.

The logarithm consists of two parts: the characteristic and the mantissa. The *characteristic* gives the basic power of 10 in whole numbers.

EXAMPLE: log 20 = 1.??
Since $10^1 = 10$ and $10^2 = 100$,

number	10	20	100
log	1.00	1.??	2.00

The characteristic is easily determined by moving the decimal in the original number so that only one digit lies to the left of the decimal. The number of places the decimal is moved equals the characteristic. The characteristic is positive if the decimal is moved to the left and negative if moved to the right.

EXAMPLE: 1,000 = 1,000. characteristic = 3
                    3 2 1

235 = 2,3,5, characteristic = 2
                2 1

.01 = .0 1 characteristic = −2
                1 2

TABLE A.2

No. x	log x			
	0	**1**	**2**	**③**
100	0000	0004	0009	0013
101	0043	0048	0052	0056
102	0086	0090	0095	0099
103	0128	0133	0137	0141
104	0170	0175	0179	0183
105	0212	0216	0220	0224
106	0253	0257	0261	0265
107	0294	0298	0302	0306
108	0334	0338	0342	0346
109	0374	0378	0382	0386
10	0000	0043	0086	0128
11	0414	0453	0492	0531
12	0792	0828	0864	(0899)
13	1139	1173	1206	1239
14	1461	1492	1523	1553

Number = *123*   Mantissa for 123

The *mantissa* gives the *fractional* distance between the powers of base 10. The mantissa is determined by the digits in the number *without* regard to the location of the decimal point and refers *only* to the relative (fractional) position of the number between powers of the base 10.

After the characteristic and the mantissa have been obtained, they can be combined to give the logarithm.

EXAMPLE: The mantissa for the number 2 equals .3010.

power of 10	10^0		10^1		10^2	
number	1	2	10	20	100	200
		.3010		.3010		.3010
log	0.0	0.3010	1	1.3010	2	2.3010

Therefore, $\log 2 = 0.3010$
$\log 20 = 1.3010$
$\log 200 = 2.3010$

The mantissa can be obtained from a "log table." (See Table P, part of which appears here as Table A.2.)

EXAMPLE: Find log of 12.3.
            characteristic   = 1.0000
            mantissa (from Table P) =   .0899
                                          _____
            log 12.3   = 1.0899

The antilog of a logarithm is determined by locating the mantissa in Table P, finding the associated number, and finally placing the decimal point as indicated by the characteristic.

Interval-data tests (i.e., t test, analysis of variance, Pearson correlation) are frequently employed, since they make efficient use of the data and are readily calculated and well understood. The disadvantage of these tests is that the individual groups of scores must be reasonably normally distributed and have similar variance. What can be done when the data are highly skewed or the variance of one group is many times that of the other?

A common way to cope with these problems is to *transform* the data so that they will meet the requirements of normality and equal variance. By algebraic manipulations ranging from simple to complex, the original scores can be "pushed, squeezed, stretched, or fluffed up" to yield the desired distribution. The most common methods are outlined below. We will also consider when such transformations are appropriate and some problems of working with transformed data.

1. *Square-root transformation.* This transformation is performed by taking the square root of the scores, and it reduces sharply scores stretching to the high end of the distribution. The square-root transformation has less effect when X is small than when X is large. (Conversely, high scores can be increased in value so that they have greater influence by squaring the scores,

Example: $X =$	2	1	-1
$\sqrt{X + 1} =$	$\sqrt{3}$	$\sqrt{2}$	$\sqrt{0}$

that is, using X^2.) Since we cannot easily calculate the square root of negative numbers, it is advisable to use $\sqrt{X + N}$ with N a constant value large enough to make all values positive.

2. *Logarithmic transformation.* Using the log of the scores serves much the same purposes as the square-root transformation. Log X affects high values more severely than does $\sqrt{X}$. The log transformation is frequently employed when the coefficient of variability (Chapter 7) is constant across groups, that is, when variability increases proportionally with increases in the mean. When X is less than 1, log X is difficult to work with, and when $X = 0$, log X is indeterminate. Therefore, it is helpful to make all values ≥ 1 by adding an appropriate constant.

Example: $X =$	2	1/2	0
$\log X + 1 =$	log 3	log 1.5	log 1

3. *Reciprocal transformation.* The reciprocal of $X = 1/X$. This transformation reduces high scores sharply (more so than the square-root or log transformations) but also increases the spread of the small scores. The relative size and influence of large and small scores is *reversed* by the reciprocal transformation. If the original scores are skewed with many low scores and few high scores, the mean will be greatly influenced by the high scores. When the reciprocal transformation is used, the mean reflects primarily the low scores.

Example: $X =$	.1	1	100
$1/X =$	10	1	.01

4. *Other transformations.* There are many other more complex and special procedures used for particular problems (such as the arc sine and probit transformations), but these are beyond the scope of the text. Problems of interpretation are especially difficult with procedures more complex than the three methods given.

5. *Ranking.* The ranking of scores is also a form of transformation, though designed to be treated with measures based on the median rather than on the mean.

TABLE B.1

	Original Scores	Transformed Scores		
		$\sqrt{X}$	log X	$-1/X$
Group 1				
	1	1.0000	0.0000	1.0000
	3	1.7321	.4771	.3333
	10	3.1623	1.0000	.1000
Σ	14	5.8944	1.4771	1.4333
$\bar{X}$	4.6667	1.9648	.4924	.4777
Median	3.0000	1.7321	.4771	.3333
$\bar{X}$ Transformed Back		3.8604	3.1090	2.0933
Median Transformed Back		3.0000	3.0000	3.0000
Group 2				
	25	5.0000	1.3979	.0400
	100	10.0000	2.0000	.0100
	400	20.0000	2.6021	.0025
Σ	525	35.0000	6.0000	.0525
$\bar{X}$	175.0000	11.6666	2.0000	.0175
Median	100.0000	10.0000	2.0000	.0100
$\bar{X}$ Transformed Back		136.1096	100.0000	57.1428
Median Transformed Back		100.0000	100.0000	100.0000
SD_1	4.7258	1.0998	.5002	.4671
SD_2	198.4313	7.6376	.6012	.0198

The effect of these transformations on two sets of scores is given in Table B.1. This table should make clear the following points about skewed distributions that stretch out toward the high side:

1. The transformations serve to equalize the variances of the two groups. For the table data the log transformation is most suitable, which shows the effectiveness of the log transformation when the coefficient of variability remains fairly constant as the means increase (for group 1, $CV_1 = 1.01$; for group 2, $CV_2 = 1.13$).
2. The $\bar{X}$ transformed back (i.e., the $\bar{X}$ of the transformed scores transformed back to the original scale) shows that the influence of high scores is reduced most by $1/X$, less with log X, and least with $\sqrt{X}$.
3. The score that is the median of the transformed scores is the same score that is the median of the original (untransformed) scores.
4. With the square-root and log transformations the transformed-back score for the $\bar{X}$ is closer to the original median than to the original mean.

The transformed-back $\overline{X}$ for the reciprocal transformation is shifted from the "high" side of the median to the "low" side.

In general, if the scores are skewed, the mean and median are not the same. If the transformation normalizes the scores, the mean and median then coincide. However, since the subject with the middle score (the median) is the same before and after transformation, the net effect of normalizing scores is to make the mean move over to the median. Therefore, a test based on the *mean* of the transformed data (e.g., t test) is essentially equivalent to a test based on the *median* of the original data (e.g., White test). The greater efficiency or sensitivity of the t test compared to the White test in this situation results from the normality and equal variance *imposed* on the data by the transformations. The transformation is not taken into account by the t test, and if it were, the two tests would be equally sensitive, since both would be using the same information (the difference between medians relative to the spread of scores about each median).

But, if transforming scores leads to no real advantage in terms of sensitivity, why bother? In the case of the t test and single-factor analysis of variance, it is probably just as effective (and certainly more direct) to use the appropriate ranking test, particularly when sample sizes are small. When sample sizes are large, it may be easier to transform the data and use a t test rather than a ranking test. The Pearson correlation and the associated regression analysis with transformed data are preferred to the relatively limited information obtainable from the rank-order correlation. Many variables that have a curvilinear relationship and are therefore unsuited to a Pearson correlation measure have a linear relationship when one or both variables are subject to a transformation. Such a linear (transformed) relationship is usually easier to comprehend and interpret than the nonlinear relationship of the original scores. Score transformations that permit the use of higher-order analysis of variance procedures are particularly helpful, since ranking tests generally do not permit measures of interaction. In analysis of variance it is the scores *within* the cells that should be normally distributed and have equal variance.

An important point, easily overlooked in discussions of transforming scores, is that conclusions based on statistical analysis of transformed scores apply to the transformed data and may or *may not* apply to the original data. Thus we may have differences or correlations between sets of transformed data that do not appear in the original data. This problem is greatest with regard to interaction—which makes the problem all the worse, since testing for interaction is often the major reason for using transformed data. Graphing the scores in Table B.1 shows the nature of the difficulty. (See Figure B.1.)

Suppose each score is the mean of a group. The striking interaction seen with the nontransformed data is greatly reduced with $\sqrt{X}$, virtually disappears with log X, and appears in *reversed* form with $1/X$. (Note that the curves

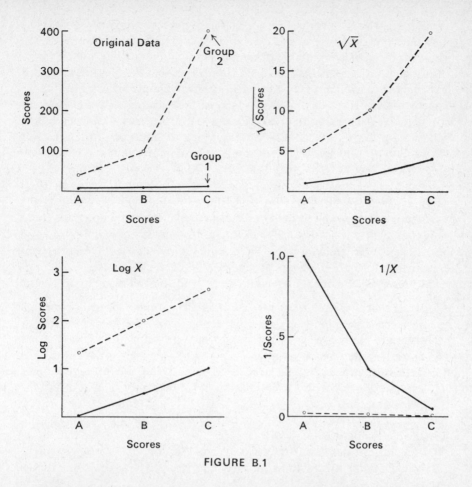

FIGURE B.1

become virtually linear with the log transformation.) Thus we can often make interactions come or go by using one or another transformation. The problem is that interaction is measured by the slope of the data lines, and stretching or compressing the scale by a transformation markedly affects the slope. The loss of slope information when scores are ranked is so great that interaction tests are virtually impossible. Differential effects upon the slope tend to be greatest when the overall level of the groups is different. Therefore, great care must be taken in the interpretation of interactions found with transformed data.

The basic problem with using transformed scores is that of interpreting the situation. Both the author and the reader of a study must keep in mind that (for example) the significant differences reported are between the mean $\sqrt{}$ scores (but not necessarily between the mean scores), and the correlation is between grades and log family income. How does one interpret (1/ anxiety level) or, worse yet, $\sqrt{}$ socioeconomic level? There is little point in transforming the data to make it amenable to a particular analysis if at the same time the situation is rendered virtually unintelligible in terms of the behavior or dependent variable that was originally measured.

For example, suppose the scores of group 1 in Table B.1 are time scores for a rat running to food in a runway or a student solving problems. How do we want to present these data? What value of the "average" would be most representative? Is the mean of the original scores too greatly affected by the high (deviant) score? If so, should we reverse the influence of the high and low scores with a reciprocal transformation or compress the high scores by using $\sqrt{X}$ or log X? The final decision depends on which measure gives a reasonable or meaningful picture of the data at the same time that it reshapes the data. In place of a complex transformation that perfectly adjusts the distributions, it may well be better to use a less efficient but simpler and more readily understood measure. For example, the square-root or log running time is reasonable and commonly used and is more serviceable than an unfamiliar transformation, chosen solely for a particular analysis.

The above discussion can be summarized with the following conclusions:

1. If the scores are highly skewed and the groups have markedly unequal variance, a ranking test should be tried. If the sample size is large or there are many groups, it may be easier to transform the data and use a t test or analysis of variance.
2. Transformations may be needed to permit the use of higher-order analysis of variance or determination of the Pearson correlation and the regression line.
3. The simplest and most understandable transformation appropriate to the data should be used.
4. Conclusions based on transformed data may or may not apply to the original data. Interactions are particularly sensitive to transformations.

test construction: reliability, validity, and test-item analysis

Reliability

Reliability is the measure of the consistency of a test, that is, the ability of a test to give the same score for a subject on repeated testing. Reliability is difficult to achieve, and unreliability arises from three main sources:

1. Context unreliability, which is due to the use of a poor sample—there may be too few items in the sample or they may not be representative of the attribute being tested.
2. Temporal unreliability, which reflects instability of the test over time—it may be highly influenced by temporary or external factors.
3. Scoring unreliability, which arises from inconsistencies in scoring. Scoring unreliability tends to be an appreciable problem in essay tests but is relatively small in objective tests.

Reliability is measured as a correlation ($r_{\text{rel.}}$) between two tests given to a sample. Typical procedures include:

1. Test-retest method. The same test is given twice. This method is not commonly used for paper-and-pencil tests. The interval between tests is critical: Temporal unreliability will enter if large intervals are used and short intervals permit memory of the first test to be a factor. The test-retest method provides a measure of stability but not of internal consistency.
2. Parallel or equivalent forms. Two versions of the test are administered. This is a commonly used approach and is applicable to both speed and

power tests.* Constructing two forms of the test, however, can be a problem.

3. Split-half or odd-even method. The scores on odd-numbered items are treated as a subtest and are compared to the scores on even-numbered items. This provides a measure of internal consistency and is frequently used for academic and power tests but is not suitable for speed tests. The odd/even form gives no measure of temporal instability, that is, changes during the test session.

Since odd-even reliability (r_{oe}) is based on two tests that are each one-half the length of the total test, the reliability of the total test should be higher than r_{oe}. On the basis of the obtained correlation (r_{oe}) between the subtests, the reliability of the total test (r_{tt}) is given by a formula known as the Spearman-Brown formula:

$$r_{tt} = \frac{2r_{oe}}{1 + r_{oe}} \qquad (C.1)$$

where $r_{\text{rel.}} = r_{tt}$
If $r_{oe} = .6$, then the reliability of the total test is:

$$r_{tt} = \frac{2(.6)}{1 + .6} = \frac{1.2}{1.6} = .75 \qquad (C.2)$$

The general form of the Spearman-Brown formula permits calculation of r_{tt} from the original test reliability (r_o) when the total test is either longer or shorter than the original test.

$$r_{\text{rel.}} = r_{tt} = \frac{Kr_o}{1 + (K - 1)r_o} \qquad (C.3)$$

where $K = \dfrac{\text{number of items in } tt}{\text{number of items in } o}$

$$r_{tt} > r_o \text{ when } K > 1$$

and

$$r_{tt} < r_o \text{ when } K < 1$$

Therefore, when the ratio of the total test length to the original test length—K in formula (C.3)—equals 2, we obtain formula (C.1). Note that formula (C.3) indicates the *decrease* in reliability if the new test is shorter than the original. If everything else is equal, the reliability of the test increases as the number of items increases.

An obvious restriction on increasing the internal consistency and reliability of tests is the difficulty in obtaining suitable additional items—this is usually more of a problem in personality tests than in academic tests. A further

---

* Speed tests are so designed that no subject or examinee has sufficient time to attempt all items. The test items are so simple that all would be answered correctly if enough time were available. Therefore, the number of correct items = the number of items attempted. Power tests provide sufficient time for the subject to attempt all items, but the items are difficult or of varying difficulty, and not all are likely to be answered correctly.

factor restricting the length of tests is the subjects' available time and energy. Responses gained from inattentive subjects have little value for judging either the subjects or the test.

The Spearman-Brown formula may also be used to indicate the proportional increase (or decrease) in test length (K) needed to change r_o to a given level of r_{tt}:

$$K = \frac{r_{tt}(1 - r_o)}{r_o(1 - r_{tt})} \tag{C.4}$$

Therefore, changing r_o from .6 so that r_{tt} will equal .75 would involve:

$$K = \frac{.75(1 - .6)}{.6(1 - .75)} = \frac{.75(.4)}{.6(.25)} = 2.0 \tag{C.5}$$

which agrees with the calculations in formula (C.2).

Since all tests are somewhat unreliable, the obtained test scores will often be somewhat higher or lower than the "true" score or score that would be obtained if the test were perfectly reliable. The *standard error of measurement* (s_e) is the standard deviation of obtained test scores about the "true" score. The value of s_e is an inverse function of $r_{rel.}$:

$$s_e = s\sqrt{1 - r_{rel.}} \tag{C.6}$$

where s = SD of the test scores
$r_{rel.}$ = coefficient of reliability

For a perfectly reliable test ($r_{rel.} = 1.0$), $s_e = 0$, and all obtained scores would be "true" ones.

Validity
Validity is a measure of the degree to which a test serves the purpose for which it is intended—that is, "does what it is supposed to do." The problem of measuring validity is a considerable one.

Validity can be judged in logical terms as *content*, or *construct*, validity— that is, whether the test appears to be an adequate sample of the desired trait or aspect or, for achievement tests, of the content or objectives of the course. Construct validity is a frequently employed measure of validity but is not as suitable as a direct, objective measure.

The best but most difficult measure of validity is a *validity coefficient* (r_v) based on the correlation with some external measure (*experimental* or *predictive* validity). The problem here is to arrive at a suitable reference or *criterion* measure. For example, how should the validity of an IQ test be determined—by reference to future school grades? achievement in adult life?

fame? fortune? A specific test (e.g., IQ) may have satisfactory validity with regard to a specific criterion (e.g., academic performance), but unless the validity is tested with regard to other criteria (e.g., adult or nonacademic achievement), viewing the test as having greater generality (e.g., with regard to "intelligence") involves a change from experimental to logical, or construct, validity.

The criterion may be a direct measure of the variable being predicted—for example, the validity of a truck-driver selection test is measured by the drivers' subsequent driving performance.* For more complex situations and behaviors (e.g., personality traits) multiple correlations and factor-analysis procedures are often employed to bring together several criterion variables.

The *correction for attenuation* is sometimes used to estimate what the validity of the test would be if the criterion measure were perfectly reliable. The corrected coefficient (r_{v_c}) is readily calculated from the obtained value of r_v and the reliability coefficient of the *criterion* variable ($r_{\text{rel.}}$):

$$r_{v_c} = \frac{r_v}{\sqrt{r_{\text{rel.}}}} \qquad\qquad (C.7)$$

When validity coefficients are reported as correlations between a measure and a criterion variable (as in personality tests), it is important to give the reliability of the *measure* to permit proper interpretation of r_v. For example, if r_v is low due to low $r_{\text{rel.}}$, r_v would improve with improvements in the measure's reliability, but if $r_{\text{rel.}}$ is already high, then the particular criterion may be a poor one.

Test-Item Analysis

Often a questionnaire needs to be developed or perfected in order to measure a particular variable. Suppose, for example, we want a measure that discriminates, or distinguishes, between the higher-scoring and lower-scoring subjects—which in a classroom situation usually means between the better and poorer students.

A common procedure for developing a test, or questionnaire, is to determine how efficiently each item or each question distinguishes between the better and poorer students. The subjects are divided into "better" and "poorer" groups on the basis of either some external measure or, more often, the total grade on the examination or test. The information afforded by this division is particularly useful as an aid in the elimination of ambiguous or inefficient items when constructing multiple-choice exams. Items on which the better and poorer students do equally well (usually those that all students answer correctly or incorrectly) do not serve to distinguish between the two

---

* This, however, is easier said than done. How is performance measured—by trip time? by accident rate? by gasoline consumption? by willingness to work for a low wage?

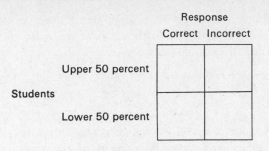

FIGURE C.1

groups. In general, very simple or very difficult items tend not to discriminate well, the best items being those of moderate difficulty (answered correctly by about 50 percent of the total group). It is important that the difficult aspect of a question be relevant to the course. For example, using large and/or obscure words in a history question may make it more difficult, but it would no longer be a test of knowledge of history.

The usual procedure for an item analysis will be briefly outlined here and one of the possible measures given. Many problems arising with regard to item analysis and test construction are beyond the scope of this text and are discussed more fully in specialized sources (e.g., Guilford 1954, Wood 1961).

The first step is to grade each test to arrive at a total test grade for each subject. The subjects are then divided into two groups on the basis of their final score (e.g., the upper and lower 50 percent) or on the basis of an external ability or achievement measure, such as IQ or prior grades. We can then analyze each test item in terms of the percentage of students in the upper group and in the lower group who answered the item correctly. This will result in a fourfold table for each item, as shown in Figure C.1.

If the categories correct/incorrect and upper 50 percent/lower 50 percent are taken as essentially nominal, the χ^2 statistic is appropriate for testing whether the high group does significantly better than the low group on a given item. However, our primary concern is not with testing the difference between the two groups but rather with determining how well the student's final grade can be predicted on the basis of his performance on a given item or how well the score on the item can be predicted on the basis of his total grade. Such a prediction calls for a measure of the relationship between the two categories or variables, and the appropriate measure for this situation is the phi coefficient (r_ϕ), which was briefly discussed in Chapter 12. The contingency coefficient (C) would be the appropriate measure if we had divided the students into more than two categories (i.e., high, medium, low grades). For a 2 × 2 table (with $df = 1$) the phi coefficient is preferable to C, since it has a range of 0 to 1.00 (when based on two groups of equal size) and can be related to the Pearson coefficient.

In this situation, the better the pass/fail groups for a given item correspond with the overall upper/lower division, the higher is the r_ϕ value and the better the total score can be predicted on the basis of that item.

The formula for the phi coefficient is:

$$r_\phi = \sqrt{\frac{\chi^2}{N}} \qquad \text{(C.8)}$$

The probability value of r_ϕ is given by the probability value of the associated χ^2. Since the upper and lower 50 percent of the subjects contain equal numbers of subjects, the calculations for r_ϕ can be markedly simplified when using the percentage of correct responses by each group as the measure:

$$r_\phi = \frac{P_{U_c} - P_{L_c}}{2\sqrt{(P)(1 - P)}} \qquad \text{(C.9)}$$

where P_{U_c} = proportion of upper group correct
P_{L_c} = proportion of lower group correct
P = proportion of total group correct

The calculation of r_ϕ can be further simplified by using Table G, which provides the appropriate r_ϕ value on the basis of the proportion of subjects in each of the two groups having a correct score for a given item. As long as the two groups contain an equal number of subjects, the exact place where the division is made is not critical to the calculation of r_ϕ. Although a common procedure is to use the upper and lower halves of the group (dividing at the median), the highest 25 percent may be compared to the lowest 25 percent or the A students may be compared to the B students. Whatever the division, r_ϕ is a measure of the degree to which the two groups respond differently to a given item or question.

The item analysis is carried out by determining the r_ϕ values for each question and then ranking the questions in terms of their efficiency (from high to low r_ϕ value). The final version of the test should consist of those items having a sufficiently high value of r_ϕ to be worth retaining (e.g., the "best" 50 items from an original group of 100 items). The test may be rerun with additional or rewritten items and the analysis repeated to refine the test further.

In order for the item analysis to be useful, the sample size has to be reasonably large (i.e., > 50), since small sample sizes lead to highly unreliable measures. Sample sizes of 100 or 200 divided into high/low groups also permit easy calculation of the percentage correct for each group. Since very good test items generally are difficult to find, the main value of an item analysis is the elimination of poor items.

The phi coefficient is only one of the correlation measures that can be used in an item analysis. Under certain conditions the point-biserial correlation

(r_{pb}) or either of two measures not discussed in the text—the biserial correlation (r_b) and the tetrachoric correlation (r_t)—may have some advantages over the phi coefficient. In general, the *absolute* level of the correlation is a function of the distribution of subjects or scores underlying the division into upper/lower and correct/incorrect and the correlation measure employed. However, the rank ordering of the items from best to worst is the same or very similar regardless of the particular measure used. The phi coefficient was selected for illustrative purposes in this section because it has been previously covered in the text and is very easy to estimate using Table G.

The ability of a particular test item to discriminate between the high-scoring and low-scoring subjects is not the only measure of the value of an item. In the case of a test designed to measure anxiety level, schizophrenic tendencies, or motivation, the discriminating ability of the items is obviously of primary importance. With the usual classroom examination the test items are presumably designed not merely to discriminate between better and poorer students but also to determine the overall level of understanding achieved by the class. This last point is important, since the elimination of items on the basis of an item analysis can cause a problem by reducing the content validity of the test. Important points or areas may be overlooked unless certain of the less efficient items are retained.

An item covering an important point in the course would fail to discriminate among the students if they all answered it correctly or if they all answered it incorrectly. Therefore, if *all* the students fully mastered the course material, the test items on the final examination would fail to discriminate among the students, and the level of performance on the test would indicate that all students did well. Similarly, there may be little value in discriminating among students when the entire class does very poorly, though it is important to know the overall level of performance of the class. Item analysis may, then, be seen as being suitable for selecting the most efficient questions to test the substantive material of a course.

Objective Tests and Essay Tests

There are two main types of classroom tests: essay examinations, which are relatively easy to write but time consuming to grade, and objective tests, which are difficult to generate but are rapidly graded. The choice between these two approaches is not clear-cut. In general, well-constructed objective tests (which generally, but *not* necessarily, use multiple-choice items) are superior in several respects to the standard essay exam.

Objective tests involve a minimum of scorer unreliability and great ease of scoring (which is particularly helpful when large numbers of subjects are involved). The achievement of adequate content sampling with objective exams (by using a sufficient range and number of items) is always a problem but is, at least, reasonably amenable to investigation.

Essay tests typically have low scorer reliability, both among scorers and within the same scorer. Literary skill and clarity of expression tend to be confounded in scoring the content (though such skill and clarity may well be valid measures of a valuable ability), and scoring is time consuming and difficult. Essays may have considerable apparent (formal) validity, but the low reliability makes the experimental validity of the test difficult to establish. In general, the drawbacks of essay exams increase as the vagueness of the question and length of the expected answer increase. Specific essay questions requiring explicit and fairly short answers can approach objective tests in efficiency and utility.

Multiple-choice tests tend to increase in reliability as the number of items increases. Reliability can also be improved by increasing the number of alternatives for each item *if* the alternatives are *equally* attractive or equally likely to be selected. (Given equally attractive items, each question acts as $K - 1$ two-choice items where K equals the number of alternatives; that is, a four-choice item equals $4 - 1 = 3$ two-choice items.) A considerable problem in constructing multiple-choice items is to obtain suitable alternatives—a test-wise subject can often score well on multiple-choice tests simply by eliminating obviously unlikely alternatives.

Poor items reduce the reliability and validity of the tests and often confuse the student, wasting time and reducing the number of items that can be administered. One way of minimizing the problem of poor alternatives and facilitating the writing of items is to use a two-choice item (*not* true-false, which is about the worst possible test). Two-alternative items are usually readily constructed, since one is the correct response and the other is the most important or readily confused incorrect response. For example:

$$\sqrt{2.5} \text{ is:} \quad \begin{array}{l} \text{(a) .50} \\ \text{(b) 1.58} \end{array}$$

Other alternatives in this question would be clearly incorrect and fail to provide useful feedback. With two-choice items the student is instructed to select the *better* of the two items. Two-choice questions can be conveniently obtained by retaining the correct answer and best alternative of available four- or five-choice items. Everything else being equal, more two-choice items than four-choice items are required for equal discrimination of better from poorer students, since chance performance would be correct 50 percent of the time. In practice, however, not very many more good two-choice items are needed to equal the efficiency of a given number of less-than-perfect four-choice items. A final score corrected for guessing can be based on the assumption that each incorrect answer indicates a "lucky guess" on another item. A chance score would therefore be 50 percent correct.

corrected score = total number − (2 × number incorrect)
                 = number correct − number incorrect

writing a scientific report:
a very brief guide

After a study or experiment has been designed, the data collected, analyzed and interpreted, the last step is to make a final report. Although colleagues may be informed of interesting results by an informal discussion or talk, the permanent and definitive statement of the findings is nearly always in the form of a journal article, monograph, or book.

Journal articles are written in a very rigid form and style, which may appear to be excessively restricting. The basic organization is much the same from one journal to the next, and the form employed by journals published by the American Psychological Association (A.P.A.) is taken as representative. Specific questions relating to the organization or style for a given journal may be answered by referring to the publication manual used by that journal or, more simply, by copying the style shown in recent articles published in that journal.

The preparation of a paper is described in the form that is used for submitting an article to a journal for publication. The basic features are:

I. Use A.P.A. journal style. The manuscript is to be typewritten, double spaced, with at least a one-inch margin on all sides. Use complete sentences throughout (except in the title, which may be a clause). Avoid telegraphese.
II. The report is in sections.
   A. Title—an abstract of the abstract giving the independent and dependent variables and, when necessary, the subjects. Be brief and avoid unnecessary words such as "A study of..." Use "key" words to help identify the paper.
   B. Abstract or summary—a brief description of the rationale and of

the method, results and discussion sections. Do not repeat the title. The abstract is on a separate page with title and author's name on top. The maximum length is 100 to 120 words. When a summary is used instead of an abstract, it comes after the discussion section and is labeled "Summary."

C. Introduction—a brief review of past work in the area and an explanation of the reasons for doing the study. Give the reader confidence by showing that you have done your "homework" and know the relevant and, at least, recent literature.

D. Method—the reader must be given enough information to be able to replicate the experiment or study on the basis of this section. Do not include unnecessary or irrelevant details.

    1. Subjects. Describe the relevant characteristics of the subjects— age, sex, weight, IQ, species, or whatever. Indicate how many subjects were used and how they were selected.

    2. Apparatus. Describe in sufficient detail (using a sketch or photograph if necessary), so that equivalent apparatus can be reconstructed or purchased. Point out specific features of the apparatus (including the stimuli) that are critical to the study.

    3. Procedure. Describe how the subjects were assigned to groups. Give the step-by-step procedure.

E. Results—show the main results in terms of scores, graphs, or tables. Point out the important aspects of each graph or table (i.e., what the reader should pay attention to). All aspects of the data that are raised in the discussion section should be first noted in this section. Show results of statistical tests.

F. Discussion—relate the results to the introduction. Do you have the answer to the original question? Interpret the data fully. What new questions are raised? Relevant studies not previously mentioned in the introduction may be referred to here.

G. References—list alphabetically by author and chronologically when several works by the same author appear. Only the sources cited in the text should be included. The references start on a separate page.

H. Tables and figures—show each one on a separate page. Table captions are part of the table. All figure captions are together on a separate page preceding the figures.

A paper is published in a journal in order to reach those individuals who are most interested in the subject area. Therefore the title of the article must contain sufficient information to catch the attention of the appropriate readers and stop them long enough to study at least the abstract. The abstract, in turn, must then provide a sufficient basis for the reader to decide whether or not it is worth spending time on the rest of the article. The title and abstract are also of special importance, since many abstracting or indexing services use only the title or only the title and the abstract. Therefore an article having a poor title and abstract is not going to reach those who should read it, and the findings in the study will likely be neglected or overlooked and their potential value not realized.

Using a formally organized paper permits the reader to locate easily any given item of information. For example, questions concerning the exact nature of the subjects or specifications of the apparatus can readily be answered by referring to the appropriate portion of the paper. Similarly, it is a simple matter to go from the discussion section back to the results section to look again at a specific table or graph. Certain other features of the paper, such as wide margins or starting the abstract or references on separate pages, greatly facilitate reviewing and editing the paper as well as preparing it for publication.

At first glance, the organization of the paper appears to be a very cumbersome one that does not tell the story of the experiment as well as you might do in an informal manner. But keep in mind that *no one* is expected to read the article in order from start to finish. The typical sequence is first to read the title and the abstract, then to glance briefly at the introduction section to see if the situation being dealt with is the one the reader expects. This is followed by a quick check of the figures and tables and a glance at the discussion section, which gives the main findings and conclusions. The discussion section might be reread or specific points looked for in the method and results sections. The opportunity to skip around while reading an article is one of the main advantages of having a formal and understood organization.

A brief paper on "Traumatic Compression" gives an example of the appropriate style and is used to illustrate the various points of the outline:

TAMING EFFECTS OF TRAUMATIC
COMPRESSION OF THE HOUSE FLY

Arnold Stroop

Wisdom State University

Flies, after compression, were significantly and appreciably less tense and more quiescent than those in a control group. These differences appeared to be long lasting. Taming was assumed to result from physiological changes in the Ss.

TAMING EFFECTS OF TRAUMATIC
COMPRESSION OF THE HOUSE FLY

Arnold Stroop

Wisdom State University

Attempts to tame the house fly (<u>Musca domestica</u>)
using conditioning techniques (Sturge,1953), punishment (Stevens,1937; Smith,1962), or manipulation of
early experience (Smedley,1965a) have met with a
consistent lack of success. Since traumatic compression is frequently employed with the fly, the
effectiveness of this procedure on taming was tested.
The term "taming" is defined behaviorally as occurring
when <u>S</u> makes no attempt to either flee or attack upon
being touched by <u>E</u>'s finger.

Method

<u>Subjects.</u> The <u>S</u>s were 24 randomly selected flies
from a population maintained in the animal room of the
W.S.U. psychology laboratory.

<u>Apparatus.</u> Compression was applied with a Swanson
Model 28-B swatter, having a 20 gm., 100 x 150 cm.
perforated polyethylene head on a flexible 14-gauge
twisted-wire shaft. All work was carried out on a
stainless-steel counter top.

<u>Procedure.</u> The <u>S</u>s were randomly assigned to the two
groups and were handled one at a time. All <u>S</u>s were
tested within a one-hour period on each of two suc-
cessive days. Each <u>S</u> was confined within a bell jar
and placed on the counter over a drop of honey. When
<u>S</u> alighted on or near the honey, the jar was carefully
removed. The experimental group was traumatically
compressed with a sharp blow. The swatter at impact
had a measured velocity of 5.13 m./sec. The control
group was given sham compression by having the blow
land just to the side of <u>S</u>.

Tameness in terms of the response to being touched
with a finger was measured on a 10-point scale with
10 = totally tame. Three observers made the
judgments independently. Tameness was tested
immediately after compression and 24 hours later.

Results

Interobserver reliability was .98, and the mean of the ratings for each S was used. The mean scores for the two groups for the immediate and 24-hour tests are shown in Table I. Two Ss in the experimental group

----

Insert Table I about here

----

were not measured, since they could not be located after the experimental manipulation. The experimental group was significantly and appreciably more tame than the control group on the immediate test ($t = 6.35$; $p < .01$; rm $> .80$) and on the 24-hour test ($t = 5.01$; $p < .01$; rm $> .80$). The 24-hour scores were negligibly different from the immediate scores for each group ($t < 1.0$; $p > .20$; rm $< .20$). All of the control Ss, but none of the experimental Ss, ate and behaved normally after compression.

Discussion

Traumatic compression appears to be a suitable means of inducing tameness in the house fly. The 24-hour scores indicated that the tameness is relatively long lasting. The effectiveness of this procedure is presumably due to the actual contact, since the sham group experienced the same handling and feeding as well as the same acoustic stimulation. The impaired behavior of the experimental Ss suggests that physiological processes influenced by the compression (Smedley,1965b) are probably important in the observed tameness.

In a related series of studies (Spock,1963), traumatic compression (though of a relatively reduced intensity) has been found useful in taming immature primates.

References

Smedley, A. Enrichment of early experience in Musca
  domestica. Journal of Ephemeral Psychology, 1965a,
  23, 117-128.
Smedley, A. Physiology and behavior change. In A.
  Simpson (Ed.), Psychophysiology. New York: McGraw
  House, 1965b.
Smith, A. Counterconditioning of the escape response
  in the house fly. Animalistic Behavior, 1962, 37,
  216-219.
Spock, A. Child care and survival. Quarterly Journal
  of Mental Health, 1963, 23, 8-12.
Stevens, A. Effect of shock on the fly. Physiological Processes, 1937, 12, 83-85. (Psychoagronomy
  Abstracts, 1939, 9, No. 634)
Sturge, A. Operant conditioning and anxiety reduction
  in three species of insects. Journal of the Experimental Control of Behavior, 1953, 4, 189-196. Cited
  by A. Sweeney, Backwaters of behavioural research.
  London: Maudley and Carter, 1960. P. 103.

TABLE I
Mean Immediate and 24-Hour Tameness Scores and Size of
the Traumatically Compressed (T.C.) and Control Groups

	N	Immediate	24 Hour
T.C.	10	9.3	9.7
Control	12	1.8	2.0

random and restricted random presentation of experimental sequences

Many experiments require that the subject be presented with a sequence of stimuli or treatments. How can these sequences be best arranged? We certainly cannot make the assignments by whim or by deciding ourselves what a "good" sequence would be, since inadvertent and unnoticed biases or patterns would be almost impossible to avoid. The subject's behavior could possibly be affected by these patterned sequences as well as by the experimental variable. A "cleaner" and more objective procedure, therefore, is required.

For example, suppose we wish to train a young child to select a red box when both a red and a blue box are placed in front of him. Candy in the correct box is the reward. How should the boxes be presented to the child? If the red box is always (or usually) on the right side, the child might choose the box on the right side without paying attention to color. If the red box is alternately placed on the right and left side, the child might learn to alternate his choice without reference to color. A sequence of positions that has no visible pattern to which the child can respond is required, so that correct responses can be made consistently on the basis of color only.

A commonly used approach is to derive the necessary sequence from a table of random numbers (using, for example, odd and even numbers for the right and left side, respectively). However, if we select a section of the table randomly (by the "eyes-closed, finger-point" method), we may find annoying problems such as too many sixes in a row or an undesirable alternation of odd and even numbers. If this occurs in practice, we have to reject the first selection and try again until we locate a "good-looking" random sequence. We usually do not want a truly random sequence, since all possible sequences and combinations should occur in a sufficiently long random series. What is

247

needed instead is a reasonably nonpredictable or nonsystematic sequence, so that inappropriate behavior on the part of the subject (e.g., alternating between the box on the left and the one on the right) would not be reinforced. Obviously, once a particular section of the random table is chosen as being "good," it can no longer be considered to be random. Such selected sequences are sometimes referred to as "haphazard assignment" or "constrained randomization"; unfortunately, however, the nonrandom features are generally *not* specified.

A reasonable approach is to use specified restrictions on a random sequence. Tables for the binary (two alternatives such as $+/-$ or right/left) and the decimal (digits 0 to 9 or 00 to 99) situations are available.

The binary (Gellerman) series (Table S) has the following specifications (Gellerman 1933):

1. In each block of ten trials, each event or alternative (R, L) appears five times.
2. In each block of five trials, there are three of one event and two of the other.
3. There are no more than three of one event in a row.
4. Blocks can be combined in any order as long as the three-in-a-row maximum is maintained.
5. Inappropriate responding in a discrimination task such as random choices or single- or double-position alternation of choices will have a 50 percent chance of being correct.

We can use the Gellerman series for the red box/blue box study. Selecting sequence 4 (based on the day of the month to avoid indecision about which order to select) leads to the following order of right (R) left (L) sides for the red box:

side: R R L R L L R R L L
trial: 1 2 3 4 5 6 7 8 9 10

If 20 trials were required, we could continue to sequence 26 in the next column. (Note that sequence 20 cannot be followed by 42, since there would be four L placements in a row.)

Suppose another experiment requires the child to select the red box from among six boxes: red (1), orange (2), yellow (3), green (4), blue (5), and violet (6). Here the Gellerman series will not work, but a restricted randomization of the digits 0 to 9 (H. Friedman 1966) can be used in similar fashion.

The decimal series (Table R) has the following specifications:

1. Each two-digit number appears once.
2. In each column each digit appears no more than twice in the tens position and no more than twice in the units position.
3. In each block of 50 numbers each digit appears five times in the tens position and five times in the units position.

4. No more than three consecutive digits (ascending or descending) appear in the tens or units positions within a column or when going from the end of one column to the start of an adjacent column.
5. Double-digit numbers, such as 55, do not appear more than twice in a column.

This series can be conveniently used to generate a great variety of sequences by starting with any of the columns and reading up or down the columns in any order within a block or within a set of columns. Additional sets can be easily obtained by transposing the tens and units columns in a set or by adding a constant to each number in a set. The properties of the tens and units positions are identical, and each block can also be used as a restricted randomization of the numbers 0 to 9 with each digit appearing ten times in the block:

Obtaining sequences of between 20 and 50 items with Table R can be facilitated by dividing the table numbers by the sequence size (or the next largest multiple of 10) and using the *remainder*. For example, to produce a sequence of 20 items, dividing the table values by 20 leads to the remainder numbers shown in Table E.1. This procedure allows all or most of the table values to be used in an assignment. If the sequence size is an ungainly number such as 17, it is easiest to use the next largest multiple of 10 (i.e., 20) as the divisor.

TABLE E.1

Table Value	Remainder after Dividing by 20
19	19
28	8
55	15
80	0 = 20

Selecting the units part of column one for our six-box example, we obtain the following sequence, with each position being used once in a block of six trials:

Numbers: 2 7 2 9 8 5 1 4 7 3
Position (L to R): 2   5 1 4   3 6

For those situations in which an unrestricted random series is required, the last two digits in the columnal numbers on a telephone directory page can generally be used. Computer-generated tables of random numbers meeting stringent definitions of "randomness" are available in many advanced texts.

studies based on a single subject ($N=1$)

Assume we have trained a rat to press a bar in a Skinner box for food and want to test the effect of loud noise on his behavior. We turn on the noise, and the rate of bar pressing sharply decreases. Does this clearly show the effect of noise? Not necessarily; perhaps the rat was tired or satiated or otherwise unwilling to press the bar. We now turn off the noise—the bar-pressing rate returns to the original level. This looks good, so we try the noise again—and the rate drops. The results so far are fairly convincing; but to be on the safe side, we try each condition (quiet and noise) twice again, and the rate changes as before. Now we are convinced that noise lowers the rate of bar pressing—for this rat at least. But would we expect similar results with other rats? And are the results so clear-cut that we have no need to use statistical analysis? More to the point, is statistical analysis inappropriate when we test only 1 subject instead of a group of subjects? The following discussion will show that not only are statistical tests appropriate when $N = 1$, but in the preceding example we used a standard statistical test (though somewhat informally).

The use and value of studies based on single subjects has been well documented (Dukes 1965). In general, these studies fall into four overlapping categories:

1. The testing of unique or rare subjects—such as a home-raised chimpanzee, an individual with a particular brain lesion, or the first captive Martian. Many studies in clinical psychology and psychiatry fall in this category.

2. The study of capacity phenomena, that is, trying to determine if creature A can do X or if X is possible. Examples of this type of problem are: Are heart transplants possible? Can opossums distinguish colors?

251

3. Studies of a single subject with no concern for the representativeness of that subject and no intention to generalize to the population. Testing a coin for fairness is a common version of this situation.

4. Problem-centered research in which the primary concern is with studying the phenomenon or process with little interest in the particular subject employed. Essentially, the subject is considered to be adequately representative. The nonsense-syllable-learning studies of Ebbinghaus are good examples in this category.

Analysis of Single-Subject Data

1. Qualitative versus quantitative data. Certain situations (particularly in tests of capacity) provide data that are *qualitatively* different from ongoing behavior, and no statistical tests need be used. For example, if you observe Dumbo rising into the air just once, the question of the possibility of flying elephants is settled. Neither repeated flights by one individual nor a squadron of airborne pachyderms is required to be convincing. However, when observations are in the form of *quantitative* differences from prior behavior, it is necessary to show that the subject's performance is not reasonably attributable to chance, and a statistical analysis is required.

2. Categorical data. Measurements are often in the form of categorical data such as correct/incorrect answers or right/left turns. Testing the fairness of a coin on the basis of head/tail categories using either a binomial or χ^2 test was discussed in Chapters 1 and 2. For example, in a test dealing with capacity phenomena, a binomial test was used for the correct/incorrect responses of individual opossums ($N = 2$ in this study) selecting a particular color in pairs of colored stimuli (Friedman 1967).

3. Score data. This type of data is appropriate to the Skinner-box noise experiment introduced at the start of this section. The specific question is whether or not the bar-pressing rates under the quiet condition (Q) are consistently higher than those under the noise condition (N). In other words, would we expect the high- and low-rate sessions to coincide with the quiet and noise sessions? The test for the overall difference in rate between quiet and noisy conditions is the White test (cf. Edgington 1967). With all four noise sessions having lower rates than the quiet sessions, $T = 10$ and $p < .05$ with $r_m > .71$. The null hypothesis in this case refers to identical *treatment* effects for a single subject, not identical *mean* or *median* effects for groups of subjects. The sequence of treatments given should be random (or restricted random) rather than simple alternation. (See Supplement E.)

In similar fashion if three or more experimental treatments are presented in random order, the Kruskal-Wallis test is appropriate for the scores.

For a well-trained subject whose overall rate remains relatively constant over treatments, the data for the Skinner-box quiet/noisy experiment might be in the form shown in Figure F.1A. Long-term satiation or fatigue effects

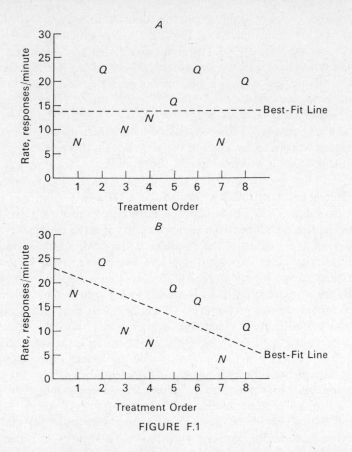

FIGURE F.1

can cause an overall decline in rate, as shown in Figure F.1B, while practice effects can lead to a general increase in rate over trials. A standard White test based on the ranks of the rate will suffice for Figure F.1A but not Figure F.1B. If the cumulative or sequential effect on the scores is relatively linear, a best-fit line can be calculated and the scores treated as deviations from this line (ranging from a large negative deviation to a large positive deviation). The White test can then be applied to these deviations. The net effect of using deviations from the best-fit linear scores is to eliminate the long-term trend and permit the difference in scores under the two treatments to be seen more clearly. As shown in Figure F.1B, the consistently high rate for Q and low rate for N, relative to the long-term trend given by the best-fit line, are very clear. In Figure F.1A, with relatively constant response rates over time, the best-fit line has a 0 slope and makes no contribution to the analysis. Curvilinear sequential effects can sometimes be made linear by the appropriate transformation (see Supplement B).

We can now see more clearly the nature of the informal procedure used in the Skinner-box quiet/noisy example. The repeated presentation of quiet and noisy conditions was required to produce confidence in the conclusion. The point at which we concluded that the effect was convincing was determined informally or nonobjectively. This point could also be determined with the aid of the White test, which operates with $N = 1$ just as it does in any other situation. It provides *objective* measures of p and r_m which aid in deciding (and convincing others) that the effect is "really there." The interpretation of quantitative data necessarily involves a statistical analysis, and an objective and explicit analysis is more reliable and informative than a subjective one. As pointed out many times in the text though, this analysis is only one step in the process of making sense of the situation.

Interpreting and Generalizing Single-Subject Data

Since the subject in $N = 1$ studies is usually tested repeatedly, the problem of contamination by sequential effects when two or more treatments are used is a serious one. Logically, the problem is the same for any study using repeated measures (see Chapter 4, topic 9), except with $N = 1$ the order of presentation cannot be counterbalanced across subjects, and the subject is generally tested more often than is typical in a group design. Random (as opposed to alternating) presentation of treatments is necessary to minimize some of the more likely sequential, or carry-over, effects. Preparing or pretraining a subject will serve to minimize warm-up, practice, and learning effects. Testing a second subject with the treatment order changed, or counterbalanced, would clarify the role of sequential effects.

Is it permissible to generalize the findings from a single-subject study when no larger sample of the population is being measured? In statistical terms, a sample of $N = 1$ provides no measure of intersubject variability and $df = 0$. This is clearly the major problem with studies based on $N = 1$ and a major reason larger samples are used when possible and practical.

There is no need to generalize to a population when the total concern of the study is a particular subject; that is, the problem is whether or not *this* coin is fair. All other coins are not being considered (category 3).

In tests dealing with capacity phenomena (category 2), only one positive, or successful, case is needed to demonstrate the *possibility*. The generality of the phenomenon can be determined by future group tests or *assumed* if there is no reason to believe that the subject is atypical, or unique. A second successful subject would answer any question of the possible uniqueness of the first subject. On the basis of two opossums (or even one) that can discriminate colors, it is reasonable (and was subsequently demonstrated) that other opossums can do so under the same conditions. Dumbo, on the other hand, has special features that may well account for his ability and suggest that flying is not a common characteristic of elephants.

In similar fashion, generalizing from unique or rare subjects (category 1) is a function of the degree to which they can be assumed to be representative of similar (but unknown or unavailable) individuals. The rationale of problem-centered research (category 4) is that the subject used *is* representative of the population.

In practice the generalization to a population from either a group of subjects or a single subject always involves a nonstatistical measure of confidence that the sample is representative. As noted previously, the use of college sophomores as a sample of all people and white rats as a sample of all mammals may often be inappropriate. A sample ($N = 1$ or $N > 1$) is representative or not in terms of those characteristics that are relevant to the dependent variable and incorrect assumptions and inappropriate generalizations are possible in any study. Considerable extra care must be exercised in interpreting studies using one or very few subjects, since no satisfactory estimate is available of the range or variability of the characteristic being investigated.

The generalization of conclusions from a single subject to the population is based on more information than just that about the subject. The generalization is governed largely by the investigator's knowledge of the population, which enables him to judge the representativeness of the sample subject, and this clearly also holds for experiments in which $N > 1$.

frequency distributions, graphs, percentiles, and standard scores

Frequency Distributions

Scores or measurements obtained from many subjects or observations are often virtually impossible to interpret until they have been presented in the form of a table or graph. Before a table or graph can be constructed, the data must be organized into simple or grouped frequency distributions.

1. *Simple frequency distributions* are obtained by listing the possible scores and indicating the frequency (f) or number of times each score appears, as shown in Table G.1. The data are now organized so that all of the same scores are shown together, and the pattern of the distribution of scores can be seen.

TABLE G.1. Number of Correct Items on a Laboratory Quiz

Score	f
1	1
2	0
3	3
4	5
⋮	⋮

2. *Grouped frequency distributions* are employed when the range of scores is large and/or many scores are not represented (i.e., $f = 0$ for many scores). The pattern in such data would be difficult to comprehend. The standard procedure for such data is to select a suitable *class interval* (i) encompassing 2, 5, 10, or more scores. For example, using $i = 5$, we might have the situation given in Table G.2.

TABLE G.2. Number of Correct
Items on a Laboratory Quiz

Interval	f
1–5	3
6–10	8
11–15	12
16–20	17
⋮	⋮

After grouping, the individual scores lose their identity, and all scores within an interval are assumed to be equally distributed throughout the interval, regardless of their original value. This assumption leads to a slight error when the calculation of statistical measures (e.g., $\overline{X}$, s) is based on grouped data, but with the ready availability of calculators and computers, statistical calculations are generally based directly on the original (raw) scores.

How many intervals should be used to divide the range of scores? The "Goldilocks formula" gives the correct answer: not too many, not too few, but just right. In practice, this often works out to 7 or 9 intervals, but 10 or even 20 may be necessary. Too many intervals will give a fine but confusing picture with many intervals having very few or no scores. Too few intervals will lose too much information and obscure the pattern of the data. Your judgment of the data is the final basis for a decision.

The *interval size* (i) follows from the number of intervals used. The usual practice is to select an interval size equal to a whole integer that is an *odd* number, so that the midpoint of the interval is also a whole integer and not a fractional value. An example is given in Table G.3.

TABLE G.3

Interval	Midpoint
1–3	2
4–6	5
7–9	8
⋮	⋮

While the intervals in Table G.3 are listed as 1–3, 4–6, and so forth, there obviously can be no gap between adjacent intervals. The interval 1–3 has the stated, or *apparent*, upper *limit* of 3. Since all measurements or scores that are closer to 3 than 4 will be included within this interval, the *real* upper *limit* is 3.4999 . . . , or up to but not including 3.5. The distinction between real and apparent limits is not a problem when the measurements are rounded off to the same unit size that is used in the intervals. When the measurement is

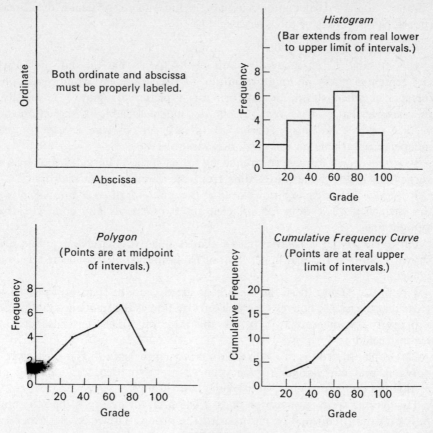

FIGURE G.1

TABLE G.4

Grade	f	Cumulative f (Cumf)
1–20	2	2
21–40	4	6
41–60	5	11
61–80	6	17
81–100	3	20
	20	

finer than the units used in the intervals, the measurements must be rounded at the time of grouping. For example, with the intervals shown in Table G.3, scores in the form of whole numbers can be placed directly in the intervals, but numbers such as 6.93 or 3.25 must first be rounded to the nearest whole number. Statistical calculations based on group data always deal with the real (not apparent) limits of the intervals.

Graphing of Data

1. Graphing of frequency distributions. Assume that the data in Table G.4 represent grades on an examination. These values can be more quickly interpreted when shown in the form of a picture or graph. Generally, the variable that is measured (i.e., the dependent variable) is shown on the vertical axis, or *ordinate*. The horizontal axis, or *abscissa*, is used for the independent variable (in this case, the class interval).

In a *frequency histogram* the frequency for each class interval is shown as a vertical bar with its width extending from the lower to upper real limits.

In a *frequency polygon* a dot placed at the midpoint of each interval shows the frequency. The dots for adjacent intervals are usually connected by straight lines.

On a *cumulative frequency curve* the cumulative frequency points are shown as dots at the real upper limit of each interval and connected as in a polygon.

Figure G.1 shows these three types of graphs for the data of Table G.4. Note that the graphs convey the same information as the frequency table, and a properly drawn graph will permit the table on which it is based to be reconstructed.

2. Graphing group or individual-performance data. The histogram, polygon, and cumulative frequency curve are also used for depi   /   comparing the performance of individuals or groups.

Histograms are often used for categorical data, for example, the presentation of group frequency or means with the groups shown on the abscissa. Histograms are particularly appropriate when there is no underlying order to the categories.

The polygon is the most commonly used type of graph. The scores or mean scores are shown on the ordinate, and conditions, times, trials, and so forth, are shown on the abscissa.

Cumulative frequency distributions are used for cumulative response data obtained in a free-operant (Skinner-box) situation. Cumulative curves are also used to show the increasing effectiveness of drugs or similar treatments with increasing dosages.

Representative graphs in these three forms are given in Figure G.2. (Note that both the ordinate and abscissa are labeled.) Whenever possible the 0 point of the dependent variable (score) should be indicated (using a break in the ordinate if necessary) to prevent misinterpretation of the overall level of the scores. This last point is illustrated in Figure G.3.

Percentiles

The relative position of a score within a distribution can be given by the rank of that score. If the group consists of 100 scores, the rank of a score is

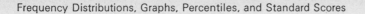

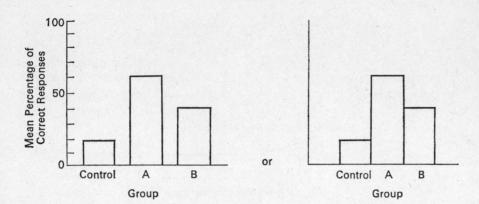

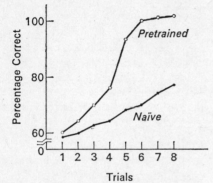

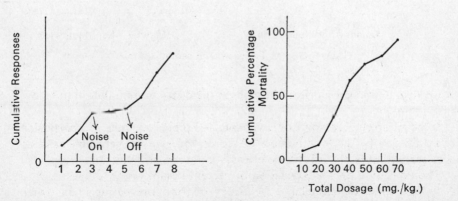

FIGURE G.2

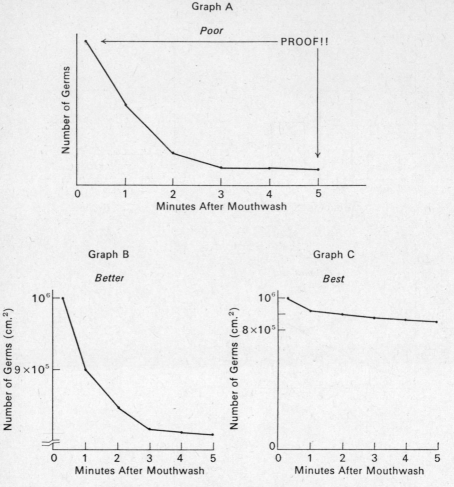

FIGURE G.3: Mouthwash Kills Mouth Germs?

equivalent to the percentage of the cases in the distribution falling at or below that score.

If the group contains more or fewer than 100 scores, the rank of a particular score divided by the total number of scores will give the **percentile rank** of that score, that is, the percentage of cases falling at or below that score.

The procedure for the calculation of percentile ranks is as follows:

1. Determine the cumulative frequency distribution, basing it on either grouped or ungrouped-data (interval size $= 1$).

2. Calculate the cumulative frequency up to the *midpoint* of each interval by subtracting $(1/2)f$, or one-half the frequency of the scores within the interval, from f of the original cumulative frequency values.

3. Convert the cumulative frequency midpoint (Cum $f_{mdpt.}$) values to cumulative proportion by dividing by N.
4. Percentile rank = cumulative proportion × 100, rounded to the nearest whole number.

The calculation of percentile ranks using the data in Table G.4 is shown in Table G.5.

TABLE G.5

Grade	f	Cum$f$	Cum$f_{mdpt.}$	Cum$f_{mdpt.}/N$	Percentile Rank
1–20	2	2	1	.050	5
21–40	4	6	4	.200	20
41–60	5	11	8.5	.425	43
61–80	6	17	14	.700	70
81–100	3	20	18.5	.925	93

where Cum$f_{mdpt.}$ = Cum$f - \left(\frac{1}{2}f\right)$

so that for the interval 41–60,

$$\text{Cum}f_{mdpt.} = 11 - \left(\frac{1}{2}5\right) = 11 - 2.5 = 8.5$$

The score associated with a given percentile rank is the *percentile* and is obtained from grouped data with the following procedure:

1. Assume that the scores are evenly distributed throughout or within the class interval.
2. Determine the cumulative frequency distribution.
3. Change the percentile being calculated to frequency (number of cases in the percentile) by multiplying the percentile rank by $N/100$.
4. Determine the position of the desired score in its interval. That is, how far into the interval is the score, or what fraction (X) of the interval lies within the percentile? If the desired score is the 3rd of the 5 scores within an interval, then X is 3/5 of the interval.
5. Multiply X by the actual size of the interval (i) to find X'. That is, X times the actual upper limit − actual lower limit = X'.
6. Add X' to the actual lower limit to obtain the score for a given percentile rank; this score equals the desired percentile.

The calculation of percentiles for the data in Table G.4 is shown in Table G.6.

TABLE G.6. The Calculation of the 50*th* Percentile (the Median), or the Score Associated with the Percentile Rank of 50

Grade	f	Cumf
1–20	2	2
21–40	4	6
	5	10 = median
41–60	5	→11
61–80	6	17
81–100	3	20
	20	

Step 3 Percentile rank $\times$ $N/100 = 50 \times 20/100 = 10$
Step 4 10 is 4/5 of the way into interval 41–60 ; $X = 4/5$
Step 5 With $i = 10, 4/5 \times i = 8 ; X' = 8$
Step 6 Actual lower limit of interval $= 40.5; 8 + 40.5 = 48.5 = 50$*th* percentile

Standard Scores: Z and T Scores

Standard scores were discussed in Chapter 7. Any score in a distribution can be expressed as a deviation from the mean (i.e., $X - \bar{X}$) in standard deviation units:

$$Z = \frac{X - \bar{X}}{\sigma} \quad \text{or} \quad \frac{X - \bar{X}}{s} \tag{G.1}$$

The use of Z scores is most suitable when the distribution is normal and serves to locate each score within the group as if the distribution had a mean equal to 0 ($\bar{X} = 0$) and SD = 1. With Z scores the relative positions of subjects within groups having different means and variability can readily be compared or measured.

A disadvantage of Z scores is that half the scores will have negative values (since they fall below the mean) and most will involve decimal values. Virtually all negative values can be avoided by increasing the mean, and decimals can be avoided by increasing the value of SD. This leads to a version of the Z score that is based on $\bar{X} = 50$ and SD = 10:

$$T = (Z \times 10) + 50 \tag{G.2}$$

A score 1.5 SD above the mean would lead to:

$$T = (1.5 \times 10) + 50 = 15 + 50 = 65 \tag{G.3}$$

Educational Testing Service reports College Board and Graduate Record Examination scores as a T score with $\bar{X} = 500$ and SD = 100:

$$T_{\text{ETS}} = (Z \times 100) + 500 = T \times 10 \qquad\qquad \text{(G.4)}$$

A score 1.5 SD above the mean would lead to:

$$T_{\text{ETS}} = (1.5 \times 100) + 500 = 150 + 500 = 650$$

or
$$T \times 10 = 65 \times 10 = 650 \qquad\qquad \text{(G.5)}$$

When the distribution of original scores is not essentially normal, Z or T scores can be misleading, but the correct position of the scores within the group can be retained by using normalized standard scores.

Normalized Standard Scores

Any distribution can be reshaped to be normal in form if each score is assigned a new value, a **normalized standard score**, which is a Z score having the same rank within a normal distribution that the score has within the distribution being reshaped. The final outcome is a normal distribution. Note that the use of Z or T scores leaves the shape of the original distribution unchanged, so that a skewed distribution would still be skewed if converted to Z scores.

The following procedure is used to obtain normalized standard scores:

1. Obtain cumulative proportion values for the midpoint of each interval ($\text{Cum}f_{\text{mdpt.}}/N$), as used in the calculation of percentile ranks shown in Table G.5.
2. Locate the cumulative proportion in the cumulative frequency ($\text{Cum}f$) columns of Table D for the normal distribution and obtain the associated Z score. For greater accuracy locate the cumulative proportion in the main body of Table D if it is not larger than .500 and obtain the associated Z score. For larger values, use $(1 - \text{cumulative proportion})$ for the entry.
3. The Z score associated with each cumulative proportion is the normalized standard score for each interval. The Z score is negative for cumulative proportions less than .50 (the mean) and positive for cumulative proportions greater than the mean.

This procedure is illustrated in Table G.7.

The use of standard scores or normalized standard scores is not a suitable means of rendering data amenable to statistical tests requiring normally distributed data. Standard scores and normalized standard scores involve essentially a total loss of information concerning the $\bar{X}$ and SD of the original data. Suitable procedures for normalizing data to use in further analysis are discussed in Supplement B on transformations.

TABLE G.7

Grade	f	Cum$f$	Cum$f_{mdpt.}$	Cum$f_{mdpt.}/N$	Normalized Standard Score
1–20	2	2	1	.050	−1.64
21–40	4	6	4	.200	−.84
41–60	5	11	8.5	.425	−.19
61–80	6	17	14	.700 (.300) [a]	.52
81–100	3	20	18.5	.925 (.075) [b]	1.44

[a] .700 = 1 − .300
[b] .925 = 1 − .075

tables

TABLE A. Table of Critical Values of the *t* Ratio [a]

df	Level of Significance for Two-Tailed Test					
	.20	.10	.05	.02	.01	.001
1	3.078	6.314	12.706	31.821	63.657	636.619
2	1.886	2.920	4.303	6.965	9.925	31.598
3	1.638	2.353	3.182	4.541	5.841	12.941
4	1.533	2.132	2.776	3.747	4.604	8.610
5	1.476	2.015	2.571	3.365	4.032	6.859
6	1.440	1.943	2.447	3.143	3.707	5.959
7	1.415	1.895	2.365	2.998	3.499	5.405
8	1.397	1.860	2.306	2.896	3.355	5.041
9	1.383	1.833	2.263	2.821	3.250	4.781
10	1.372	1.812	2.228	2.764	3.169	4.587
11	1.363	1.796	2.201	2.718	3.106	4.437
12	1.356	1.782	2.179	2.681	3.055	4.318
13	1.350	1.771	2.160	2.650	3.012	4.221
14	1.345	1.761	2.145	2.624	2.977	4.140
15	1.341	1.753	2.131	2.602	2.947	4.073
16	1.337	1.746	2.120	2.583	2.921	4.015
17	1.333	1.740	2.110	2.567	2.898	3.965
18	1.330	1.734	2.101	2.552	2.878	3.922
19	1.328	1.729	2.093	2.539	2.861	3.883
20	1.325	1.725	2.086	2.528	2.845	3.850
21	1.323	1.721	2.080	2.518	2.831	3.819
22	1.321	1.717	2.074	2.508	2.819	3.792
23	1.319	1.714	2.069	2.500	2.807	3.767
24	1.318	1.711	2.064	2.492	2.797	3.745
25	1.316	1.708	2.060	2.485	2.787	3.725
26	1.315	1.706	2.056	2.479	2.779	3.707
27	1.314	1.703	2.052	2.473	2.771	3.690
28	1.313	1.701	2.048	2.467	2.763	3.674
29	1.311	1.699	2.045	2.462	2.756	3.659
30	1.310	1.697	2.042	2.457	2.750	3.646
40	1.303	1.684	2.021	2.423	2.704	3.551
60	1.296	1.671	2.000	2.390	2.660	3.460
120	1.289	1.658	1.980	2.358	2.617	3.373
∞	1.282	1.645	1.960	2.326	2.576	3.291

[a] Each table entry is the minimum *t* value (based on a given *df*) required for significance at each probability level.

SOURCE : Table A is taken from Table III of Fisher and Yates : *Statistical Tables for Biological, Agricultural, and Medical Research* (4th ed.), published by Oliver & Boyd, Edinburgh, and by permission of the authors and publishers.

TABLE B. Simplified Table of Magnitude of Experimental Effect (r_m)

Part I

r_m as a Function of the Sample Size Needed to Detect the Effect at a Given Two-Tailed Probability Level

Sample Size	Probability				
	.20	.10	.05	.01	.001
1	.95	.99	1.00	1.00	1.00
2	.80	.90	.96	.99	1.00
3	.69	.81	.88	.96	.99
4	.61	.73	.81	.92	.97
5	.55	.67	.75	.87	.95
6	.51	.62	.71	.83	.92
7	.47	.58	.67	.80	.90
8	.44	.55	.63	.76	.87
9	.42	.52	.60	.73	.85
10	.40	.50	.58	.70	.82
11	.38	.48	.55	.68	.80
12	.36	.46	.53	.66	.78
13	.35	.44	.51	.64	.76
14	.34	.43	.50	.62	.74
15	.33	.41	.48	.60	.72
16	.32	.40	.47	.59	.71
17	.31	.39	.46	.58	.69
18	.30	.38	.44	.56	.68
19	.29	.37	.43	.55	.67
20	.28	.36	.42	.54	.65
21	.28	.35	.41	.53	.64
22	.27	.34	.40	.52	.63
23	.27	.34	.40	.51	.62
24	.26	.33	.39	.50	.61
25	.25	.32	.38	.49	.60
26	.25	.32	.37	.48	.59
27	.25	.31	.37	.47	.58
28	.24	.31	.36	.46	.57
29	.24	.30	.36	.46	.56
30	.23	.30	.35	.45	.55
40	.20	.26	.30	.39	.49
60	.17	.21	.25	.32	.41
120	.12	.15	.18	.23	.29

Part II

The Measure of Sample Size for Several Statistical Tests

Test	Sample Size
t test	df
F test	$df_{\text{error term}}$
U test or White test	$(N_1 - 1) + (N_2 - 1)$
Matched-Pair Signed-Ranks	$N - 1$
χ^2	N = total observations
Kruskal-Wallis	N = total observations
Friedman test	$N(k)$ = total observations
Binomial test	N = total observations

SOURCE: H. Friedman. A simplified table for the estimation of magnitude of experimental effect. *Psychonomic Science,* 1969, **14**(4), 193.

TABLE C. Expanded Table of Magnitude of Experimental Effect (r_m)

Part I

The Measure Equivalent to r_m for Several Inferential Tests with Values for Use in Part II

Test	r_m	S	Entry Q²	Q
t	r_{pb}	df		t
Z	r_{pb}	N		Z
F	eta	df_D	$df_N(F)$	
χ^2	C	N	χ^2	

$$r_m = \sqrt{\frac{Q^2}{Q^2 + S}}$$

Part II

r_m as a Function of the Value of the Statistic (Q or Q^2) and the Sample Size (S)

Values of t equivalent to $p = .05$ and $p = .01$ (for a two-tailed test) for each level of r_m are indicated by an asterisk and a dagger, respectively.

For each level of r_m (assuming two samples have equal size and equal variance), the equivalent difference between means in standard score units (DM/σ), and the equivalent proportion misclassified [$P(m)$] (when subjects are assigned to groups on the basis of their scores) are given at the bottom of each column. (Table follows.)

$r_m =$	.2		.25		.3		.35		.4	
s	Q^2	Q	Q^2	Q	Q^2	Q	Q^2	Q	Q^2	Q
1.	.04	.20	.07	.26	.10	.31	.14	.37	.19	.44
2.	.08	.29	.13	.37	.20	.44	.28	.53	.38	.62
3.	.13	.35	.20	.45	.30	.54	.42	.65	.57	.76
4.	.17	.41	.27	.52	.40	.63	.56	.75	.76	.87
5.	.21	.46	.33	.58	.49	.70	.70	.84	.95	.98
6.	.25	.50	.40	.63	.59	.77	.84	.92	1.14	1.07
7.	.29	.54	.47	.68	.69	.83	.98	.99	1.33	1.15
8.	.33	.58	.53	.73	.79	.89	1.12	1.06	1.52	1.23
9.	.38	.61	.60	.77	.89	.94	1.26	1.12	1.71	1.31
10.	.42	.65	.67	.82	.99	.99	1.40	1.18	1.90	1.38
11.	.46	.68	.73	.86	1.09	1.04	1.54	1.24	2.10	1.45
12.	.50	.71	.80	.89	1.19	1.09	1.68	1.29	2.29	1.51
13.	.54	.74	.87	.93	1.29	1.13	1.81	1.35	2.48	1.57
14.	.58	.76	.93	.97	1.38	1.18	1.95	1.40	2.67	1.63
15.	.62	.79	1.00	1.00	1.48	1.22	2.09	1.45	2.86	1.69
16.	.67	.82	1.07	1.03	1.58	1.26	2.23	1.49	3.05	1.75
17.	.71	.84	1.13	1.06	1.68	1.30	2.37	1.54	3.24	1.80
18.	.75	.87	1.20	1.10	1.78	1.33	2.51	1.59	3.43	1.85
19.	.79	.89	1.27	1.13	1.88	1.37	2.65	1.63	3.62	1.90
20.	.83	.91	1.33	1.15	1.98	1.41	2.79	1.67	3.81	1.95
21.	.88	.94	1.40	1.18	2.08	1.44	2.93	1.71	4.00	2.00
22.	.92	.96	1.47	1.21	2.18	1.48	3.07	1.75	4.19	2.05
23.	.96	.98	1.53	1.24	2.27	1.51	3.21	1.79	4.38	2.09*
24.	1.00	1.00	1.60	1.26	2.37	1.54	3.35	1.83	4.57	2.14

25.	1.04	1.02	1.67	1.29	2.47	1.57	3.49	1.87	4.76	2.18
26.	1.08	1.04	1.73	1.32	2.57	1.60	3.63	1.91	4.95	2.23
27.	1.13	1.06	1.80	1.34	2.67	1.63	3.77	1.94	5.14	2.27
28.	1.17	1.08	1.87	1.37	2.77	1.66	3.91	1.98	5.33	2.31
29.	1.21	1.10	1.93	1.39	2.87	1.69	4.05	2.01	5.52	2.35
30.	1.25	1.12	2.00	1.41	2.97	1.72	4.19	2.05*	5.71	2.39
31.	1.29	1.14	2.07	1.44	3.07	1.75	4.33	2.08	5.90	2.43
32.	1.33	1.15	2.13	1.46	3.16	1.78	4.47	2.11	6.10	2.47
33.	1.38	1.17	2.20	1.48	3.26	1.81	4.61	2.15	6.29	2.51
34.	1.42	1.19	2.27	1.51	3.36	1.83	4.75	2.18	6.48	2.54
35.	1.46	1.21	2.33	1.53	3.46	1.86	4.89	2.21	6.67	2.58
36.	1.50	1.22	2.40	1.55	3.56	1.89	5.03	2.24	6.86	2.62
37.	1.54	1.24	2.47	1.57	3.66	1.91	5.17	2.27	7.05	2.65
38.	1.58	1.26	2.53	1.59	3.76	1.94	5.30	2.30	7.24	2.69
39.	1.63	1.27	2.60	1.61	3.86	1.96	5.44	2.33	7.43	2.73†
40.	1.67	1.29	2.67	1.63	3.96	1.99	5.58	2.36	7.62	2.76
45.	1.87	1.37	3.00	1.73	4.45	2.11*	6.28	2.51	8.57	2.93
50.	2.08	1.44	3.33	1.83	4.95	2.22	6.98	2.64	9.52	3.09
60.	2.50	1.58	4.00	2.00*	5.93	2.44	8.38	2.89†	11.43	3.38
70.	2.92	1.71	4.67	2.16	6.92	2.63	9.77	3.13	13.33	3.65
80.	3.33	1.83	5.33	2.31	7.91	2.81†	11.17	3.34	15.24	3.90
90.	3.75	1.94	6.00	2.45	8.90	2.98	12.56	3.54	17.14	4.14
100.	4.17	2.04*	6.67	2.58	9.89	3.14	13.96	3.74	19.05	4.36
125.	5.21	2.28	8.33	2.89†	12.36	3.52	17.45	4.18	23.81	4.88
150.	6.25	2.50	10.00	3.16	14.84	3.85	20.94	4.58	28.57	5.35
200.	8.33	2.89†	13.33	3.65	19.78	4.45	27.92	5.28	38.10	6.17
DM/σ	.41		.51		.63		.75		.87	
$P(m)$		.42		.40		.38		.35		.33

$r_m =$	.45		.5		.55		.6	
s	Q^2	Q	Q^2	Q	Q^2	Q	Q^2	Q
1.	.25	.50	.33	.58	.43	.66	.56	.75
2.	.51	.71	.67	.82	.87	.93	1.13	1.06
3.	.76	.87	1.00	1.00	1.30	1.14	1.69	1.30
4.	1.02	1.01	1.33	1.15	1.73	1.32	2.25	1.50
5.	1.27	1.13	1.67	1.29	2.17	1.47	2.81	1.68
6.	1.52	1.23	2.00	1.41	2.60	1.61	3.38	1.84
7.	1.78	1.33	2.33	1.53	3.04	1.74	3.94	1.98
8.	2.03	1.43	2.67	1.63	3.47	1.86	4.50	2.12
9.	2.29	1.51	3.00	1.73	3.90	1.98	5.06	2.25
10.	2.54	1.59	3.33	1.83	4.34	2.08	5.63	2.37*
11.	2.79	1.67	3.67	1.91	4.77	2.18	6.19	2.49
12.	3.05	1.75	4.00	2.00	5.20	2.28*	6.75	2.60
13.	3.30	1.82	4.33	2.08	5.64	2.37	7.31	2.70
14.	3.55	1.89	4.67	2.16*	6.07	2.46	7.88	2.81
15.	3.81	1.95	5.00	2.24	6.51	2.55	8.44	2.91
16.	4.06	2.02	5.33	2.31	6.94	2.63	9.00	3.00†
17.	4.32	2.08	5.67	2.38	7.37	2.71	9.56	3.09
18.	4.57	2.14*	6.00	2.45	7.81	2.79	10.13	3.18
19.	4.82	2.20	6.33	2.52	8.24	2.87†	10.69	3.27
20.	5.08	2.25	6.67	2.58	8.67	2.95	11.25	3.35
21.	5.33	2.31	7.00	2.65	9.12	3.02	11.81	3.44
22.	5.59	2.36	7.33	2.71	9.54	3.09	12.38	3.52
23.	5.84	2.42	7.67	2.77	9.97	3.16	12.94	3.60
24.	6.09	2.47	8.00	2.83†	10.41	3.23	13.50	3.67
25.	6.35	2.52	8.33	2.89	10.84	3.29	14.06	3.75

26.	6.60	2.57	8.67	2.94	11.28	3.36	14.63	3.82
27.	6.86	2.62	9.00	3.00	11.71	3.42	15.19	3.90
28.	7.11	2.67	9.33	3.06	12.14	3.48	15.75	3.97
29.	7.36	2.71	9.67	3.11	12.58	3.55	16.31	4.04
30.	7.62	2.76†	10.00	3.16	13.01	3.61	16.88	4.11
31.	7.87	2.81	10.33	3.21	13.44	3.67	17.44	4.18
32.	8.13	2.85	10.67	3.27	13.88	3.73	18.00	4.24
33.	8.38	2.89	11.00	3.32	14.31	3.78	18.56	4.31
34.	8.63	2.94	11.33	3.37	14.75	3.84	19.13	4.37
35.	8.89	2.98	11.67	3.42	15.18	3.90	19.69	4.44
36.	9.14	3.02	12.00	3.46	15.61	3.95	20.25	4.50
37.	9.39	3.07	12.33	3.51	16.05	4.01	20.81	4.56
38.	9.65	3.11	12.67	3.56	16.48	4.06	21.38	4.62
39.	9.90	3.15	13.00	3.61	16.91	4.11	21.94	4.68
40.	10.16	3.19	13.33	3.65	17.35	4.17	22.50	4.74
45.	11.43	3.38	15.00	3.87	19.52	4.42	25.31	5.03
50.	12.70	3.56	16.67	4.08	21.68	4.66	28.13	5.30
60.	15.24	3.90	20.00	4.47	26.02	5.10	33.75	5.81
70.	17.77	4.22	23.33	4.83	30.36	5.51	39.38	6.27
80.	20.31	4.51	26.67	5.16	34.70	5.89	45.00	6.71
90.	22.85	4.78	30.00	5.48	39.03	6.25	50.63	7.12
100.	25.39	5.04	33.33	5.77	43.37	6.59	56.25	7.50
125.	31.74	5.63	41.67	6.45	54.21	7.36	70.31	8.39
150.	38.09	6.17	50.00	7.07	65.05	8.07	84.38	9.19
200.	50.78	7.13	66.67	8.16	86.74	9.31	112.50	10.61
DM/σ	1.01		1.15		1.32		1.50	
$P(m)$	.31		.28		.26		.23	

$r_m =$	.65		.7		.75		.8	
s	Q^2	Q	Q^2	Q	Q^2	Q	Q^2	Q
1.	.73	.86	.96	.98	1.29	1.13	1.78	1.33
2.	1.46	1.21	1.92	1.39	2.57	1.60	3.56	1.89
3.	2.19	1.48	2.88	1.70	3.86	1.96	5.33	2.31
4.	2.93	1.71	3.84	1.96	5.14	2.27	7.11	2.67
5.	3.66	1.91	4.80	2.19	6.45	2.54	8.89	2.98*
6.	4.39	2.10	5.76	2.40	7.71	2.78*	10.67	3.27
7.	5.12	2.26	6.73	2.59*	9.00	3.00	12.44	3.53†
8.	5.85	2.42*	7.69	2.77	10.29	3.21	14.22	3.77
9.	6.58	2.57	8.65	2.94	11.57	3.40†	16.00	4.00
10.	7.32	2.70	9.61	3.10	12.86	3.59	17.78	4.22
11.	8.05	2.84	10.57	3.25†	14.14	3.76	19.56	4.42
12.	8.78	2.96	11.53	3.40	15.43	3.93	21.33	4.62
13.	9.51	3.08†	12.49	3.53	16.71	4.09	23.11	4.81
14.	10.24	3.20	13.45	3.67	18.00	4.24	24.89	4.99
15.	10.97	3.31	14.41	3.80	19.29	4.39	26.67	5.16
16.	11.71	3.42	15.37	3.92	20.57	4.54	28.44	5.33
17.	12.44	3.53	16.33	4.04	21.86	4.68	30.22	5.50
18.	13.17	3.63	17.29	4.16	23.14	4.81	32.00	5.66
19.	13.90	3.73	18.25	4.27	24.43	4.94	33.78	5.81
20.	14.63	3.83	19.22	4.38	25.71	5.07	35.56	5.96
21.	15.36	3.92	20.18	4.49	27.00	5.20	37.33	6.11
22.	16.10	4.01	21.14	4.60	28.29	5.32	39.11	6.25
23.	16.83	4.10	22.10	4.70	29.57	5.44	40.89	6.39
24.	17.56	4.19	23.06	4.80	30.86	5.55	42.67	6.53
25.	18.29	4.28	24.02	4.90	32.14	5.67	44.44	6.67

26.	19.02	4.36	24.98	5.00	33.43	5.78	46.22	6.80
27.	19.75	4.44	25.94	5.09	34.71	5.89	48.00	6.95
28.	20.48	4.53	26.90	5.19	36.00	6.00	49.78	7.06
29.	21.22	4.61	27.86	5.28	37.29	6.11	51.56	7.18
30.	21.95	4.68	28.82	5.37	38.57	6.21	53.33	7.30
31.	22.68	4.76	29.78	5.46	39.86	6.31	55.11	7.42
32.	23.41	4.84	30.75	5.55	41.14	6.41	56.89	7.54
33.	24.14	4.91	31.71	5.63	42.43	6.51	58.67	7.66
34.	24.87	4.99	32.67	5.72	43.71	6.61	60.44	7.77
35.	25.61	5.06	33.63	5.80	45.00	6.71	62.22	7.89
36.	26.34	5.13	34.59	5.88	46.29	6.80	64.00	8.00
37.	27.07	5.20	35.55	5.96	47.57	6.90	65.78	8.11
38.	27.80	5.27	36.51	6.04	48.86	6.99	67.56	8.22
39.	28.53	5.34	37.47	6.12	50.14	7.08	69.33	8.33
40.	29.26	5.41	38.43	6.20	51.43	7.17	71.11	8.43
45.	32.92	5.74	43.24	6.58	57.86	7.61	80.00	8.94
50.	36.58	6.05	48.04	6.93	64.29	8.02	88.89	9.43
60.	43.90	6.63	57.65	7.59	77.14	8.78	106.67	10.33
70.	51.21	7.16	67.26	8.20	90.00	9.49	124.44	11.16
80.	58.53	7.65	76.86	8.77	102.86	10.14	142.22	11.93
90.	65.84	8.11	86.47	9.30	115.71	10.76	160.00	12.65
100.	73.16	8.55	96.08	9.80	128.57	11.34	177.78	13.33
125.	91.45	9.56	120.10	10.96	160.71	12.68	222.22	14.91
150.	109.74	10.48	144.12	12.00	192.86	13.89	266.67	16.33
200.	146.32	12.10	192.16	13.86	257.14	16.04	355.56	18.86
DM/σ		1.71		1.96		2.27		2.67
$P(m)$		.20		.16		.13		.09

SOURCE: Reproduced from H. Friedman. Magnitude of experimental effect and a table for its rapid estimation. *Psychological Bulletin*, 1968, **70**(4), 245–251, and printed by permission of the American Psychological Association.

TABLE D. Proportional Areas of the Normal Distribution[a]

Part I

Z	.00	.01	.02	.03	.04	.05	.06	.07	.08	.09	Cumulative Area	
											$Z = -3 \to 0$	$Z = 0 \to +3$
.0	.5000	.4960	.4920	.4880	.4840	.4801	.4761	.4721	.4681	.4641	.5000	.5000
.1	.4602	.4562	.4522	.4483	.4443	.4404	.4364	.4325	.4286	.4247	.4602	.5398
.2	.4207	.4168	.4129	.4090	.4052	.4013	.3974	.3936	.3897	.3859	.4207	.5793
.3	.3821	.3783	.3745	.3707	.3669	.3632	.3594	.3557	.3520	.3483	.3821	.6179
.4	.3446	.3409	.3372	.3336	.3300	.3264	.3228	.3192	.3156	.3121	.3446	.6554
.5	.3085	.3050	.3015	.2981	.2946	.2912	.2877	.2843	.2810	.2776	.3085	.6915
.6	.2743	.2709	.2676	.2643	.2611	.2578	.2546	.2514	.2483	.2451	.2743	.7257
.7	.2420	.2389	.2358	.2327	.2296	.2266	.2236	.2206	.2177	.2148	.2420	.7580
.8	.2119	.2090	.2061	.2033	.2005	.1977	.1949	.1922	.1894	.1867	.2119	.7881
.9	.1841	.1814	.1788	.1762	.1736	.1711	.1685	.1660	.1635	.1611	.1841	.8159
1.0	.1587	.1562	.1539	.1515	.1492	.1469	.1446	.1423	.1401	.1379	.1587	.8413
1.1	.1357	.1335	.1314	.1292	.1271	.1251	.1230	.1210	.1190	.1170	.1357	.8643
1.2	.1151	.1131	.1112	.1093	.1075	.1056	.1038	.1020	.1003	.0985	.1151	.8849
1.3	.0968	.0951	.0934	.0918	.0901	.0885	.0869	.0853	.0838	.0823	.0968	.9032
1.4	.0808	.0793	.0778	.0764	.0749	.0735	.0721	.0708	.0694	.0681	.0808	.9192
1.5	.0668	.0655	.0643	.0630	.0618	.0606	.0594	.0582	.0571	.0559	.0668	.9332
1.6	.0548	.0537	.0526	.0516	.0505	.0495	.0485	.0475	.0465	.0455	.0548	.9452
1.7	.0446	.0436	.0427	.0418	.0409	.0401	.0392	.0384	.0375	.0367	.0446	.9554

Z score	.00	.01	.02	.03	.04	.05	.06	.07	.08	.09	1 − Table value
1.8	.0359	.0351	.0344	.0336	.0329	.0322	.0314	.0307	.0301	.0294	.9641
1.9	.0287	.0281	.0274	.0268	.0262	.0256	.0250	.0244	.0239	.0233	.9713
2.0	.0228	.0222	.0217	.0212	.0207	.0202	.0197	.0192	.0188	.0183	.9772
2.1	.0179	.0174	.0170	.0166	.0162	.0158	.0154	.0150	.0146	.0143	.9821
2.2	.0139	.0136	.0132	.0129	.0125	.0122	.0119	.0116	.0113	.0110	.9861
2.3	.0107	.0104	.0102	.0099	.0096	.0094	.0091	.0089	.0087	.0084	.9893
2.4	.0082	.0080	.0078	.0075	.0073	.0071	.0069	.0068	.0066	.0064	.9918
2.5	.0062	.0060	.0059	.0057	.0055	.0054	.0052	.0051	.0049	.0048	.9930
2.6	.0047	.0045	.0044	.0043	.0041	.0040	.0039	.0038	.0037	.0036	.9953
2.7	.0035	.0034	.0033	.0032	.0031	.0030	.0029	.0028	.0027	.0026	.9965
2.8	.0026	.0025	.0024	.0023	.0023	.0022	.0021	.0021	.0020	.0019	.9974
2.9	.0019	.0018	.0018	.0017	.0016	.0016	.0015	.0015	.0014	.0014	.9981
3.0	.0013	.0013	.0013	.0013	.0012	.0012	.0011	.0011	.0011	.0010	.9987
	Table value										1 − Table value

Part II

Z Scores Associated with Selected Two-Tailed Probability Values

Two-tailed probability value	$p = .20$	.10	.05	.01	.001
Z score	$Z = 1.2816$	1.6449	1.9600	2.5758	3.2905

[a] The main body of the table gives the proportional area lying beyond a given Z score, that is, in the tail of the distribution. These areas are the one-tailed probabilities for values as extreme as the given value of Z.
The cumulative area columns give the proportional area lying to the left of a given Z score. For Z < 0, these values are the same as the main table values, and for Z > 0, these values are equal to 1 minus the table value.

TABLE E. Critical Values of the F Ratio for .05 (Roman), .01 (Italic), and .001 (Boldface) Levels of Significance[a]

df_D ↓ \ df_N →	1	2	3	4	5	6	8	12	24	∞
1	161	200	216	225	230	234	239	244	249	254
	4052	*4999*	*5403*	*5625*	*5724*	*5859*	*5981*	*6106*	*6234*	*6366*
	405284	**500000**	**540379**	**562500**	**576405**	**585937**	**598144**	**610667**	**623497**	**636619**
2	18.51	19.00	19.16	19.25	19.30	19.33	19.37	19.41	19.45	19.50
	98.49	*99.01*	*99.17*	*99.25*	*99.30*	*99.33*	*99.36*	*99.42*	*99.46*	*99.50*
	998.5	**999.0**	**999.2**	**999.2**	**999.3**	**999.3**	**999.4**	**999.4**	**999.5**	**999.5**
3	10.13	9.55	9.28	9.12	9.01	8.94	8.84	8.74	8.64	8.53
	34.12	*30.81*	*29.46*	*28.71*	*28.24*	*27.91*	*27.49*	*27.05*	*26.60*	*26.12*
	167.5	**148.5**	**141.1**	**137.1**	**134.6**	**132.8**	**130.6**	**128.3**	**125.9**	**123.5**
4	7.71	6.94	6.59	6.39	6.26	6.16	6.04	5.91	5.77	5.63
	21.20	*18.00*	*16.69*	*15.98*	*15.52*	*15.21*	*14.80*	*14.37*	*13.93*	*13.46*
	74.14	**61.25**	**56.18**	**53.44**	**51.71**	**50.53**	**49.00**	**47.41**	**45.77**	**44.05**
5	6.61	5.79	5.41	5.19	5.05	4.95	4.82	4.68	4.53	4.36
	16.26	*13.27*	*12.06*	*11.39*	*10.97*	*10.67*	*10.27*	*9.89*	*9.47*	*9.02*
	47.04	**36.61**	**33.20**	**31.09**	**29.75**	**28.84**	**27.64**	**26.42**	**25.14**	**23.78**
	5.99	5.14	4.76	4.53	4.39	4.28	4.15	4.00	3.84	3.67

6	*6.88* **15.75**	*7.31* **16.89**	*7.72* **17.99**	*8.10* **19.03**	*8.47* **20.03**	*8.75* **20.81**	*9.15* **21.90**	*9.78* **23.70**	*10.92* **27.00**	*13.74* **35.51**
7	*3.23* *5.65* **11.69**	*3.41* *6.07* **12.73**	*3.57* *6.47* **13.71**	*3.73* *6.84* **14.63**	*3.87* *7.19* **15.52**	*3.97* *7.46* **16.21**	*4.12* *7.85* **17.19**	*4.35* *8.45* **18.77**	*4.74* *9.55* **21.69**	*5.59* *12.25* **29.22**
8	*2.93* *4.86* **9.34**	*3.12* *5.28* **10.30**	*3.28* *5.67* **11.19**	*3.44* *6.03* **12.04**	*3.58* *6.37* **12.86**	*3.69* *6.63* **13.49**	*3.84* *7.01* **14.39**	*4.07* *7.59* **15.83**	*4.46* *8.65* **18.49**	*5.32* *11.26* **25.42**
9	*2.71* *4.37* **7.81**	*2.90* *4.73* **8.72**	*3.07* *5.11* **9.57**	*3.23* *5.47* **10.37**	*3.37* *5.80* **11.13**	*3.48* *6.06* **11.71**	*3.63* *6.42* **12.56**	*3.86* *6.99* **13.90**	*4.26* *8.02* **16.39**	*5.12* *10.56* **22.86**
10	*2.54* *3.91* **6.76**	*2.74* *4.33* **7.64**	*2.91* *4.71* **8.45**	*3.07* *5.06* **9.20**	*3.22* *5.39* **9.92**	*3.33* *5.64* **10.48**	*3.48* *5.99* **11.28**	*3.71* *6.55* **12.55**	*4.10* *7.56* **14.91**	*4.96* *10.04* **21.04**
11	*2.40* *3.60* **6.00**	*2.61* *4.02* **6.85**	*2.79* *4.40* **7.63**	*2.95* *4.74* **8.35**	*3.09* *5.07* **9.05**	*3.20* *5.32* **9.58**	*3.36* *5.67* **10.35**	*3.59* *6.22* **11.56**	*3.98* *7.20* **13.81**	*4.84* *9.65* **19.69**
12	*2.30* *3.36* **5.42**	*2.50* *3.78* **6.25**	*2.69* *4.16* **7.00**	*2.85* *4.50* **7.71**	*3.00* *4.82* **8.38**	*3.11* *5.06* **8.89**	*3.26* *5.41* **9.63**	*3.49* *5.95* **10.80**	*3.88* *6.93* **12.97**	*4.75* *9.33* **18.64**

TABLE E. Critical Values of the *F* Ratio for .05 (Roman), .01 (*Italic*), and .001 (**Boldface**) Levels of Significance (*continued*)

df_D ↓ / df_N →	1	2	3	4	5	6	8	12	24	∞
13	4.67 *9.07* **17.81**	3.80 *6.70* **12.31**	3.41 *5.74* **10.21**	3.18 *5.20* **9.07**	3.02 *4.86* **8.35**	2.92 *4.62* **7.86**	2.77 *4.30* **7.21**	2.60 *3.96* **6.52**	2.42 *3.59* **5.78**	2.21 *3.16* **4.97**
14	4.60 *8.86* **17.14**	3.74 *6.51* **11.78**	3.34 *5.56* **9.73**	3.11 *5.03* **8.62**	2.96 *4.69* **7.92**	2.85 *4.46* **7.43**	2.70 *4.14* **6.80**	2.53 *3.80* **6.13**	2.35 *3.43* **5.41**	2.13 *3.00* **4.60**
15	4.54 *8.68* **16.59**	3.68 *6.36* **11.34**	3.29 *5.42* **9.34**	3.06 *4.89* **8.25**	2.90 *4.56* **7.57**	2.79 *4.32* **7.09**	2.64 *4.00* **6.47**	2.48 *3.67* **5.81**	2.29 *3.29* **5.10**	2.07 *2.87* **4.31**
16	4.49 *8.53* **16.12**	3.63 *6.23* **10.97**	3.24 *5.29* **9.00**	3.01 *4.77* **7.94**	2.85 *4.44* **7.27**	2.74 *4.20* **6.81**	2.59 *3.89* **6.19**	2.42 *3.55* **5.55**	2.24 *3.18* **4.85**	2.01 *2.75* **4.06**
17	4.45 *8.40* **15.72**	3.59 *6.11* **10.66**	3.20 *5.18* **8.73**	2.96 *4.67* **7.68**	2.81 *4.34* **7.02**	2.70 *4.10* **6.56**	2.55 *3.79* **5.96**	2.38 *3.45* **5.32**	2.19 *3.08* **4.63**	1.96 *2.65* **3.85**
18	4.41 *8.28*	3.55 *6.01*	3.16 *5.09*	2.93 *4.58*	2.77 *4.25*	2.66 *4.01*	2.51 *3.71*	2.34 *3.37*	2.15 *3.00*	1.92 *2.57*

	15.38	10.39	8.49	7.46	6.81	6.35	5.76	5.13	4.45	3.67
19	4.38	3.52	3.13	2.90	2.74	2.63	2.48	2.31	2.11	1.88
	8.18	5.93	5.01	4.50	4.17	3.94	3.63	3.30	2.92	2.49
	15.08	**10.16**	**8.28**	**7.26**	**6.61**	**6.18**	**5.59**	**4.97**	**4.29**	**3.52**
20	4.35	3.49	3.10	2.87	2.71	2.60	2.45	2.28	2.08	1.84
	8.10	5.85	4.94	4.43	4.10	3.87	3.56	3.23	2.86	2.42
	14.82	**9.95**	**8.10**	**7.10**	**6.46**	**6.02**	**5.44**	**4.82**	**4.15**	**3.38**
21	4.32	3.47	3.07	2.84	2.68	2.57	2.42	2.25	2.05	1.81
	8.02	5.78	4.87	4.37	4.04	3.81	3.51	3.17	2.80	2.36
	14.59	**9.77**	**7.94**	**6.95**	**6.32**	**5.88**	**5.31**	**4.70**	**4.03**	**3.26**
22	4.30	3.44	3.05	2.82	2.66	2.55	2.40	2.23	2.03	1.78
	7.94	5.72	4.82	4.31	3.99	3.76	3.45	3.12	2.75	2.31
	14.38	**9.61**	**7.80**	**6.81**	**6.19**	**5.76**	**5.19**	**4.58**	**3.92**	**3.15**
23	4.28	3.42	3.03	2.80	2.64	2.53	2.38	2.20	2.00	1.76
	7.88	5.66	4.76	4.26	3.94	3.71	3.41	3.07	2.70	2.26
	14.19	**9.47**	**7.67**	**6.69**	**6.08**	**5.65**	**5.09**	**4.48**	**3.82**	**3.05**
24	4.26	3.40	3.01	2.78	2.62	2.51	2.36	2.18	1.98	1.73
	7.82	5.61	4.72	4.22	3.90	3.67	3.36	3.03	2.66	2.21
	14.03	**9.34**	**7.55**	**6.59**	**5.98**	**5.55**	**4.99**	**4.39**	**3.74**	**2.97**

TABLE E. Critical Values of the F Ratio for .05 (Roman), .01 (Italic), and .001 (Boldface) Levels of Significance (continued)

df_D ↓ \\ df_N →	1	2	3	4	5	6	8	12	24	∞
25	4.24 7.77 **13.88**	3.38 5.57 **9.22**	2.99 4.68 **7.45**	2.76 4.18 **6.49**	2.60 3.86 **5.88**	2.49 3.63 **5.46**	2.34 3.32 **4.91**	2.16 2.99 **4.31**	1.96 2.62 **3.66**	1.71 2.17 **2.89**
26	4.22 7.72 **13.74**	3.37 5.53 **9.12**	2.98 4.64 **7.36**	2.74 4.14 **6.41**	2.59 3.82 **5.80**	2.47 3.59 **5.38**	2.32 3.29 **4.83**	2.15 2.96 **4.24**	1.95 2.58 **3.59**	1.69 2.13 **2.82**
27	4.21 7.68 **13.61**	3.35 5.49 **9.02**	2.96 4.60 **7.27**	2.73 4.11 **6.33**	2.57 3.78 **5.73**	2.46 3.56 **5.31**	2.30 3.26 **4.76**	2.13 2.93 **4.17**	1.93 2.55 **3.52**	1.67 2.10 **2.75**
28	4.20 7.64 **13.50**	3.34 5.45 **8.93**	2.95 4.57 **7.19**	2.71 4.07 **6.25**	2.56 3.75 **5.66**	2.44 3.53 **5.24**	2.29 3.23 **4.69**	2.12 2.90 **4.11**	1.91 2.52 **3.46**	1.65 2.06 **2.70**
29	4.18 7.60 **13.39**	3.33 5.42 **8.85**	2.93 4.54 **7.12**	2.70 4.04 **6.19**	2.54 3.73 **5.59**	2.43 3.50 **5.18**	2.28 3.20 **4.64**	2.10 2.87 **4.05**	1.90 2.49 **3.41**	1.64 2.03 **2.64**

df_D										
30	4.17	3.32	2.92	2.69	2.53	2.42	2.27	2.09	1.89	1.62
	7.56	5.39	4.51	4.02	3.70	3.47	3.17	2.84	2.47	2.01
	13.29	**8.77**	**7.05**	**6.12**	**5.53**	**5.12**	**4.58**	**4.00**	**3.36**	**2.59**
40	4.08	3.23	2.84	2.61	2.45	2.34	2.18	2.00	1.79	1.51
	7.31	5.18	4.31	3.83	3.51	3.29	2.99	2.66	2.29	1.80
	12.61	**8.25**	**6.60**	**5.70**	**5.13**	**4.73**	**4.21**	**3.64**	**3.01**	**2.23**
60	4.00	3.15	2.76	2.52	2.37	2.25	2.10	1.92	1.70	1.39
	7.08	4.98	4.13	3.65	3.34	3.12	2.82	2.50	2.12	1.60
	11.97	**7.76**	**6.17**	**5.31**	**4.76**	**4.37**	**3.87**	**3.31**	**2.69**	**1.90**
120	3.92	3.07	2.68	2.45	2.29	2.17	2.02	1.83	1.61	1.25
	6.85	4.79	3.95	3.48	3.17	2.96	2.66	2.34	1.95	1.38
	11.38	**7.31**	**5.79**	**4.95**	**4.42**	**4.04**	**3.55**	**3.02**	**2.40**	**1.56**
∞	3.84	2.99	2.60	2.37	2.21	2.09	1.94	1.75	1.52	1.00
	6.64	4.60	3.78	3.32	3.02	2.80	2.51	2.18	1.79	1.00
	10.83	**6.91**	**5.42**	**4.62**	**4.10**	**3.74**	**3.27**	**2.74**	**2.13**	**1.00**

[a] Each table entry is the minimum value of F [based on df for the numerator (df_N) and df for the denominator (df_D)] required for significance at each probability level.

SOURCE: Table F of Q. McNemar. *Psychological Statistics.* New York: John Wiley and Sons, 1969; the table is based on Table V of Fisher and Yates: *Statistical Tables for Biological, Agricultural, and Medical Research,* published by Oliver & Boyd, Edinburgh, and by permission of the authors and publishers.

TABLE F. Critical Values of χ^2 [a]

	Level of Significance for Two-Tailed Test					
df	.20	.10	.05	.02	.01	.001
1	1.64	2.71	3.84	5.41	6.64	10.83
2	3.22	4.60	5.99	7.82	9.21	13.82
3	4.64	6.25	7.82	9.84	11.34	16.27
4	5.99	7.78	9.49	11.67	13.28	18.46
5	7.29	9.24	11.07	13.39	15.09	20.52
6	8.56	10.64	12.59	15.03	16.81	22.46
7	9.80	12.02	14.07	16.62	18.48	24.32
8	11.03	13.36	15.51	18.17	20.09	26.12
9	12.24	14.68	16.92	19.68	21.67	27.88
10	13.44	15.99	18.31	21.16	23.21	29.59
11	14.63	17.28	19.68	22.62	24.72	31.26
12	15.81	18.55	21.03	24.05	26.22	32.91
13	16.98	19.81	22.36	25.47	27.69	34.53
14	18.15	21.06	23.68	26.87	29.14	36.12
15	19.31	22.31	25.00	28.26	30.58	37.70
16	20.46	23.54	26.30	29.63	32.00	39.29
17	21.62	24.77	27.59	31.00	33.41	40.75
18	22.76	25.99	28.87	32.35	34.80	42.31
19	23.90	27.20	30.14	33.69	36.19	43.82
20	25.04	28.41	31.41	35.02	37.57	45.32
21	26.17	29.62	32.67	36.34	38.93	46.80
22	27.30	30.81	33.92	37.66	40.29	48.27
23	28.43	32.01	35.17	38.97	41.64	49.73
24	29.55	33.20	36.42	40.27	42.98	51.18
25	30.68	34.38	37.65	41.57	44.31	52.62
26	31.80	35.56	38.88	42.86	45.64	54.05
27	32.91	36.74	40.11	44.14	46.96	55.48
28	34.03	37.92	41.34	45.42	48.28	56.89
29	35.14	39.09	42.69	46.69	49.59	58.30
30	36.25	40.26	43.77	47.96	50.89	59.70

[a] For df greater than 30, the value obtained from the expression $\sqrt{2\chi^2} - \sqrt{2df - 1}$ may be used as a t ratio.

Each table entry is the minimum value of χ^2 (based on a given df) required for significance at each probability level.

SOURCE: Table F is taken from Table IV of Fisher and Yates: *Statistical Tables for Biological, Agricultural, and Medical Research* (4th ed.), published by Oliver & Boyd, Edinburgh, and by permission of the authors and publishers.

TABLE G. Values of the r_ϕ Coefficient[a]

P_{U_c} = Proportion Upper Group Correct

P_{L_c} = Proportion Lower Group Correct	0	5	10	15	20	25	30	35	40	45	50	55	60	65	70	75	80	85	90	95	100	
0	–	16	23	28	33	38	42	46	50	54	58	62	65	69	73	77	82	86	90	95	100	0
5		–	09	17	23	28	33	38	42	46	50	55	59	63	67	71	76	80	85	90	95	5
10			–	08	14	20	25	30	35	39	44	48	52	57	61	66	70	75	80	85	90	10
15				–	07	13	18	23	28	33	37	42	46	51	56	60	65	70	75	80	86	15
20					–	06	12	17	22	27	31	36	41	46	50	55	60	65	70	76	82	20
25						–	06	11	16	21	26	31	35	40	45	50	55	60	66	71	77	25
30							–	05	10	15	20	25	30	35	40	45	50	56	61	67	73	30
35								–	05	10	15	20	25	30	35	40	46	51	57	63	69	35
40									–	05	10	15	20	25	30	35	41	46	52	59	65	40
45										–	05	10	15	20	25	31	36	42	48	55	62	45
50											–	05	10	15	20	26	31	37	44	50	58	50
55												–	05	10	15	21	27	33	39	46	54	55
60													–	05	10	16	22	28	35	42	50	60
65														–	05	11	17	23	30	38	46	65
70															–	06	12	18	25	33	42	70
75																–	06	13	20	28	38	75
80																	–	07	14	23	33	80
85																		–	08	17	28	85
90																			–	09	23	90
95																				–	16	95
100																					–	100

[a] This table can be used either when there is an equal division of the sample (i.e., $N_{upper} = N_{lower}$) or there are equal-sized samples. The table is entered with the proportion correct in the upper group (P_{U_c}) and in the lower group (P_{L_c}). The value of r_ϕ, rounded to two decimal places, is given (with the decimal point omitted) in the cell at the intersection of the selected columns.

This table was recalculated to permit rounding to two digits. A larger table of r_ϕ values for P_{U_c} and P_{L_c} in unit steps is given in C. E. Jurgensen. Table for determining phi coefficients. *Psychometrika*. March, 1947. **12**(1), 17–29.

TABLE H. Critical Values for the Binomial Test When $p = q = 1/2$[a]

N	Two-Tailed Probability Values					
	.20	.10	.05	.02	.01	.001
5	0	0				
6	0	0	0			
7	1	0	0	0		
8	1	1	0	0	0	
9	2	1	1	0	0	
10	2	1	1	0	0	
11	2	2	1	1	0	0
12	3	2	2	1	1	0
13	3	3	2	1	1	0
14	4	3	2	2	1	0
15	4	3	3	2	2	1
16	4	4	3	2	2	1
17	5	4	4	3	2	1
18	5	5	4	3	3	1
19	6	5	4	4	3	2
20	6	5	5	4	3	2
21	7	6	5	4	4	2
22	7	6	5	5	4	3
23	7	7	6	5	4	3
24	8	7	6	5	5	3
25	8	7	7	6	5	4

N						
26	9	8	7		6	4
27	9	8			6	4
28	10	9	8	7	6	5
29	10	9	8	7	7	5
30		10	9	8	7	5
31	11	10	9	8	7	5
32	11		10	8	8	6
33	12	11	10	9	8	6
34	12	11	10		9	6
35	13	12	11	10	9	7
36	13	12	11	10	9	7
37	14	13	12	11	10	7
38	14	13	12	11	10	8
39	15	13	12		11	8
40	15	14	13	12	11	9
41	15	14	13	12	11	9
42	16	15	14	13	12	9
43	16	15	14	13	12	10
44	17	16	15	13	13	10
45	17	16	15	14	13	11
46	18	16	15	14	13	11
47	18	17	16	15	14	11
48	19	17	16	15	14	12
49	19	18	17	15	15	12
50	20	18	17	16	15	12

[a] X equals the number of times the less frequent of two possible events occurs. The table gives the maximum number of times X can occur in N cases for each probability level.

TABLE I. Table of Critical Values for T or T^1 (Whichever Is Smaller) for the White Test[a]

5% Critical Points of Rank Sums for a Two-Tailed Test

n_2 ↓ \ n_1 →	2	3	4	5	6	7	8	9	10	11	12	13	14	15
4			10											
5		6	11	17										
6		7	12	18	26									
7		7	13	20	27	36								
8	3	8	14	21	29	38	49							
9	3	8	15	22	31	40	51	63						
10	3	9	15	23	32	42	53	65	78					
11	4	9	16	24	34	44	55	68	81	96				
12	4	10	17	26	35	46	58	71	85	99	115			
13	4	10	18	27	37	48	60	73	88	103	119	137		
14	4	11	19	28	38	50	63	76	91	106	123	141	160	
15	4	11	20	29	40	52	65	79	94	110	127	145	164	185
16	4	12	21	31	42	54	67	82	97	114	131	150	169	
17	5	12	21	32	43	56	70	84	100	117	135	154		
18	5	13	22	33	45	58	72	87	103	121	139			
19	5	13	23	34	46	60	74	90	107	124				
20	5	14	24	35	48	62	77	93	110					
21	6	14	25	37	50	64	79	95						
22	6	15	26	38	51	66	82							
23	6	15	27	39	53	68								
24	6	16	28	40	55									
25	6	16	28	42										
26	7	17	29											
27	7	17												
28	7													

1% Critical Points of Rank Sums for a Two-Tailed Test

n_2 ↓ \ n_1 →	2	3	4	5	6	7	8	9	10	11	12	13	14	15
5			15											
6			10	16	23									
7			10	17	24	32								
8			11	17	25	34	43							
9		6	11	18	26	35	45	56						
10		6	12	19	27	37	47	58	71					
11		6	12	20	28	38	49	61	74	87				
12		7	13	21	30	40	51	63	76	90	106			
13		7	14	22	31	41	53	65	79	93	109	125		
14		7	14	22	32	43	54	67	81	96	112	129	147	
15		8	15	23	33	44	56	70	84	99	115	133	151	171
16		8	15	24	34	46	58	72	86	102	119	137	155	
17		8	16	25	36	47	60	74	89	105	122	140		
18		8	16	26	37	49	62	76	92	108	125			
19	3	9	17	27	38	50	64	78	94	111				
20	3	9	18	28	39	52	66	81	97					
21	3	9	18	29	40	53	68	83						
22	3	10	19	29	42	55	70							
23	3	10	19	30	43	57								
24	3	10	20	31	44									
25	3	11	20	32										
26	3	11	21											
27	4	11												
28	4													

0.1% Critical Points of Rank Sums for a Two-Tailed Test

n_2 ↓ \ n_1 →	3	4	5	6	7	8	9	10	11	12	13	14	15
7					28								
8				21	29	38							
9			15	22	30	40	50						
10			15	23	31	41	52	63					
11			16	23	32	42	53	65	78				
12			16	24	33	43	55	67	81	95			
13		10	17	25	34	45	56	69	83	98	114		
14		10	17	26	35	46	58	71	85	100	116	134	
15		10	18	26	36	47	60	73	87	103	119	137	156
16		11	18	27	37	49	61	75	90	105	122	140	
17		11	19	28	38	50	63	77	92	108	125		
18		11	19	29	39	51	65	79	94	111			
19		12	20	29	41	53	66	81	97				
20		12	20	30	42	54	68	83					
21	6	12	21	31	43	56	70						
22	6	13	21	32	44	57							
23	6	13	22	33	45								
24	6	13	23	34									
25	6	14	23										
26	6	14											
27	7												

[a] The table is entered with n_1 = the size of the smaller sample and n_2 = the size of the larger sample. The intersection of the selected column and row gives the maximum value for T or T^1 for the indicated probability level.

SOURCE: C. White. The use of ranks in a test of significance for comparing two treatments. *Biometrics*, March, 1952, 33–41.

TABLE J. Critical Values of T for the Matched-Pairs Signed-Ranks Test[a]

N	Level of Significance for Two-Tailed Test				N	Level of Significance for Two-Tailed Test			
	.10	.05	.02	.01		.10	.05	.02	.01
5	0	—	—	—	28	130	116	101	91
6	2	0	—	—	29	140	126	110	100
7	3	2	0	—	30	151	137	120	109
8	5	3	1	0					
9	8	5	3	1	31	163	147	130	118
10	10	8	5	3	32	175	159	140	128
					33	187	170	151	138
11	13	10	7	5	34	200	182	162	148
12	17	13	9	7	35	213	195	173	159
13	21	17	12	9					
14	25	21	15	12	36	227	208	185	171
15	30	25	19	15	37	241	221	198	182
					38	256	235.	211	194
16	35	29	23	19	39	271	249	224	207
17	41	34	27	23	40	286	264	238	220
18	47	40	32	27					
19	53	46	37	32	41	302	279	252	233
20	60	52	43	37	42	319	294	266	247
					43	336	310	281	261
21	67	58	49	42	44	353	327	296	276
22	75	65	55	48	45	371	343	312	291
23	83	73	62	54					
24	91	81	69	61	46	389	361	328	307
25	100	89	76	68	47	407	378	345	322
					48	426	396	362	339
26	110	98	84	75	49	446	415	379	355
27	119	107	92	83	50	466	434	397	373

[a] Each table entry is the maximum value of T (for $N \leq 50$) for significance at the given probability level.

SOURCE: Table J of R. P. Runyon and A. Haber. *Fundamentals of Behavioral Statistics.* Reading, Mass.: Addison-Wesley Publishing Co., 1967; based on F. Wilcoxon, S. Katti, and R. A. Wilcox. *Critical Values and Probability Levels for the Wilcoxon Rank Sum Test and the Wilcoxon Signed Rank Test.* New York: American Cyanamid Co., 1963.

TABLE K. Two-Tailed Probability Values Associated with Values of H of the Kruskal-Wallis Test

Sample Sizes						Sample Sizes				
n_1	n_2	n_3	H	p		n_1	n_2	n_3	H	p
2	1	1	2.7000	.500		4	3	2	6.4444	.008
									6.3000	.011
2	2	1	3.6000	.200					5.4444	.046
									5.4000	.051
2	2	2	4.5714	.067					4.5111	.098
			3.7143	.200					4.4444	.102
3	1	1	3.2000	.300		4	3	3	6.7455	.010
3	2	1	4.2857	.100					6.7091	.013
			3.8571	.133					5.7909	.046
									5.7273	.050
3	2	2	5.3572	.029					4.7091	.092
			4.7143	.048					4.7000	.101
			4.5000	.067						
			4.4643	.105		4	4	1	6.6667	.010
									6.1667	.022
3	3	1	5.1429	.043					4.9667	.048
			4.5714	.100					4.8667	.054
			4.0000	.129					4.1667	.082
									4.0667	.102
3	3	2	6.2500	.011						
			5.3611	.032		4	4	2	7.0364	.006
			5.1389	.061					6.8727	.011
			4.5556	.100					5.4545	.046
			4.2500	.121					5.2364	.052
									4.5545	.098
3	3	3	7.2000	.004					4.4455	.103
			6.4889	.011						
			5.6889	.029		4	4	3	7.1439	.010
			5.6000	.050					7.1364	.011
			5.0667	.086					5.5985	.049
			4.6222	.100					5.5758	.051
									4.5455	.099
4	1	1	3.5714	.200					4.4773	.102
4	2	1	4.8214	.057						
			4.5000	.076		4	4	4	7.6538	.008
			4.0179	.114					7.5385	.011
									5.6923	.049
4	2	2	6.0000	.014					5.6538	.054
			5.3333	.033					4.6539	.097
			5.1250	.052					4.5001	.104
			4.4583	.100						
			4.1667	.105		5	1	1	3.8571	.143
4	3	1	5.8333	.021		5	2	1	5.2500	.036
			5.2083	.050					5.0000	.048
			5.0000	.057					4.4500	.071
			4.0556	.093					4.2000	.095
			3.8889	.129					4.0500	.119

Sample Sizes					Sample Sizes				
n_1	n_2	n_3	H	p	n_1	n_2	n_3	H	p
5	2	2	6.5333	.008				5.6308	.050
			6.1333	.013				4.5487	.099
			5.1600	.034				4.5231	.103
			5.0400	.056					
			4.3733	.090	5	4	4	7.7604	.009
			4.2933	.122				7.7440	.011
								5.6571	.049
5	3	1	6.4000	.012				5.6176	.050
			4.9600	.048				4.6187	.100
			4.8711	.052				4.5527	.102
			4.0178	.095					
			3.8400	.123	5	5	1	7.3091	.009
								6.8364	.011
5	3	2	6.9091	.009				5.1273	.046
			6.8218	.010				4.9091	.053
			5.2509	.049				4.1091	.086
			5.1055	.052				4.0364	.105
			4.6509	.091					
			4.4945	.101	5	5	2	7.3385	.010
								7.2692	.010
5	3	3	7.0788	.009				5.3385	.047
			6.9818	.011				5.2462	.051
			5.6485	.049				4.6231	.097
			5.5152	.051				4.5077	.100
			4.5333	.097					
			4.4121	.109	5	5	3	7.5780	.010
								7.5429	.010
5	4	1	6.9545	.008				5.7055	.046
			6.8400	.011				5.6264	.051
			4.9855	.044				4.5451	.100
			4.8600	.056				4.5363	.102
			3.9873	.098					
			3.9600	.102	5	5	4	7.8229	.010
								7.7914	.010
5	4	2	7.2045	.009				5.6657	.049
			7.1182	.010				5.6429	.050
			5.2727	.049				4.5229	.099
			5.2682	.050				4.5200	.101
			4.5409	.098					
			4.5182	.101	5	5	5	8.0000	.009
								7.9800	.010
5	4	3	7.4449	.010				5.7800	.049
			7.3949	.011				5.6600	.051
			5.6564	.049				4.5600	.100
								4.5000	.102

SOURCE: Table O of S. Siegel. *Nonparametric Statistics for the Behavioral Sciences.* New York: McGraw-Hill Book Co., 1956; based on W. H. Kruskal and W. A. Wallis. Use of ranks in one-criterion variance analysis. *Journal of the American Statistical Association.* 1952, **47**, 614–617, and Errata, *Journal of the American Statistical Association.* 1953, **48**, 910.

TABLE L. Critical Values of χ_r^2 of the Friedman Test[a]

| | $k = 3$ | | | |
| | Level of Significance for a Two-Tailed Test | | | |
N	.10	.05	.01	.001
3		6.0		
4	6.0	6.5	8.0	
5	5.2	6.4	8.4	10.0
6	5.3	7.0	9.0	12.0
7	5.4	7.1	8.8	12.2
8	5.2	6.2	9.0	12.2
9	5.5	6.2	8.6	12.6

| | $k = 4$ | | | |
| | Level of Significance for a Two-Tailed Test | | | |
N	.10	.05	.01	.001
2		6.0		
3	6.6	7.4	9.0	
4	6.3	7.8	9.6	11.1

[a] Each table entry is the minimum value of χ_r^2 required for significance at a given probability level.

All χ_r^2 values shown are cut off, not rounded, to one decimal place.

SOURCE: Adapted from Table N of S. Siegel. *Nonparametric Statistics for the Behavioral Sciences.* New York: McGraw-Hill Book Co., 1956; based on M. Friedman. The use of ranks to avoid the assumption of normality implicit in the analysis of variance. *Journal of the American Statistical Association,* 1937, **32**, 088–089.

TABLE M. Critical Values of r_s, the Rank-Order Correlation Coefficient[a]

	Level of Significance for a Two-Tailed Test			
N	.10	.05	.02	.01
4	1.000	—	—	—
5	0.900	1.000	1.000	—
6	0.829	0.886	0.943	1.000
7	0.714	0.786	0.893	0.929
8	0.643	0.738	0.833	0.881
9	0.600	0.683	0.783	0.833
10	0.564	0.648	0.745	0.794
11	0.520	0.620	0.735	0.791
12	0.506	0.591	0.712	0.787
13	0.475	0.566	0.671	0.744
14	0.456	0.544	0.645	0.719
15	0.441	0.524	0.622	0.688
16	0.425	0.506	0.601	0.665
17	0.412	0.490	0.582	0.644
18	0.399	0.475	0.564	0.625
19	0.388	0.462	0.548	0.607
20	0.377	0.450	0.534	0.591
21	0.368	0.438	0.520	0.576
22	0.359	0.428	0.508	0.562
23	0.351	0.418	0.496	0.549
24	0.343	0.409	0.485	0.537
25	0.336	0.400	0.475	0.526
26	0.329	0.392	0.465	0.515
27	0.323	0.384	0.456	0.505
28	0.317	0.377	0.448	0.496
29	0.311	0.370	0.440	0.487
30	0.305	0.364	0.432	0.478

[a] Each table entry is the minimum value of r_s required for significance at a given probability level for a given sample size (N).

SOURCE : Based on E. G. Olds. The 5 percent significance levels of sums of squares of rank differences and a correction. *Annals of Mathematical Statistics.* 1949, **20**, 117–118, and E. G. Olds. Distribution of the sum of squares of rank differences for small numbers of individuals. *Annals of Mathematical Statistics,* 1938, **9,** 133–148.

TABLE N. Critical Values of *r*, the Pearson Correlation Coefficient[a]

df = N − 2	Level of Significance for a Two-Tailed Test				
	.1	.05	.02	.01	.001
1	.98769	.99692	.999507	.999877	.9999988
2	.90000	.95000	.98000	.990000	.99900
3	.8054	.8783	.93433	.95873	.99116
4	.7293	.8114	.8822	.91720	.97406
5	.6694	.7545	.8329	.8745	.95074
6	.6215	.7067	.7887	.8343	.92493
7	.5822	.6664	.7498	.7977	.8982
8	.5494	.6319	.7155	.7646	.8721
9	.5214	.6021	.6851	.7348	.8471
10	.4973	.5760	.6581	.7079	.8233
11	.4762	.5529	.6339	.6835	.8010
12	.4575	.5324	.6120	.6614	.7800
13	.4409	.5139	.5923	.6411	.7603
14	.4259	.4973	.5742	.6226	.7420
15	.4124	.4821	.5577	.6055	.7246
16	.4000	.4683	.5425	.5897	.7084
17	.3887	.4555	.5285	.5751	.6932
18	.3783	.4438	.5155	.5614	.6787
19	.3687	.4329	.5034	.5487	.6652
20	.3598	.4227	.4921	.5368	.6524
25	.3233	.3809	.4451	.4869	.5974
30	.2960	.3494	.4093	.4487	.5541
35	.2746	.3246	.3810	.4182	.5189
40	.2573	.3044	.3578	.3932	.4896
45	.2428	.2875	.3384	.3721	.4648
50	.2306	.2732	.3218	.3541	.4433
60	.2108	.2500	.2948	.3248	.4078
70	.1954	.2319	.2737	.3017	.3799
80	.1829	.2172	.2565	.2830	.3568
90	.1726	.2050	.2122	.2673	.3376
100	.1638	.1946	.2301	.2540	.3211

[a] Each table entry is the minimum value of *r* required for significance at a given probability level and a given sample size.

SOURCE: Table N is taken from Table VI of R. A. Fisher and F. Yates. *Statistical Tables for Biological, Agricultural, and Medical Research* (4th ed.)., published by Oliver & Boyd, Edinburgh, and by permission of the authors and publishers.

TABLE O. Squares and Square Roots of Numbers from 1 to 1,000

N	N^2	$\sqrt{N}$	N	N^2	$\sqrt{N}$
1	1	1.0000	44	19 36	6.6332
2	4	1.4142	45	20 25	6.7082
3	9	1.7321			
4	16	2.0000	46	21 16	6.7823
5	25	2.2361	47	22 09	6.8557
			48	23 04	6.9282
6	36	2.4495	49	24 01	7.0000
7	49	2.6458	50	25 00	7.0711
8	64	2.8284			
9	81 ·	3.0000	51	26 01	7.1414
10	1 00	3.1623	52	27 04	7.2111
			53	28 09	7.2801
11	1 21	3.3166	54	29 16	7.3485
12	1 44	3.4641	55	30 25	7.4162
13	1 69	3.6056			
14	1 96	3.7417	56	31 36	7.4833
15	2 25	3.8730	57	32 49	7.5498
			58	33 64	7.6158
16	2 56	4.0000	59	34 81	7.6811
17	2 89	4.1231	60	36 00	7.7460
18	3 24	4.2426			
19	3 61	4.3589	61	37 21	7.8102
20	4 00	4.4721	62	38 44	7.8740
			63	39 69	7.9373
21	4 41	4.5826	64	40 96	8.0000
22	4 84	4.6904	65	42 25	8.0623
23	5 29	4.7958			
24	5 76	4.8990	66	43 56	8.1240
25	6 25	5.0000	67	44 89	8.1854
			68	46 24	8.2462
26	6 76	5.0990	69	47 61	8.3066
27	7 29	5.1962	70	49 00	8.3666
28	7 84	5.2915			
29	8 41	5.3852	71	50 41	8.4261
30	9 00	5.4772	72	51 84	8.4853
			73	53 29	8.5440
31	9 61	5.5678	74	54 76	8.6023
32	10 24	5.6569	75	56 25	8.6603
33	10 89	5.7446			
34	11 56	5.8310	76	57 76	8.7178
35	12 25	5.9161	77	59 29	8.7750
			78	60 84	8.8318
36	12 96	6.0000	79	62 41	8.8882
37	13 69	6.0828	80	64 00	8.9443
38	14 44	6.1644			
39	15 21	6.2450	81	65 61	9.0000
40	16 00	6.3246	82	67 24	9.0554
			83	68 89	9.1104
41	16 81	6.4031	84	70 56	9.1652
42	17 64	6.4807	85	72 25	9.2195
43	18 49	6.5574			

N	N²	√N	N	N²	√N
86	73 96	9.2736	129	1 66 41	11.3578
87	75 69	9.3274	130	1 69 00	11.4018
88	77 44	9.3808			
89	79 21	9.4340	131	1 71 61	11.4455
90	81 00	9.4868	132	1 74 24	11.4891
			133	1 76 89	11.5326
91	82 81	9.5394	134	1 79 56	11.5758
92	84 64	9.5917	135	1 82 25	11.6190
93	86 49	9.6437			
94	88 36	9.6954	136	1 84 69	11.6619
95	90 25	9.7468	137	1 87 69	11.7047
			138	1 90 44	11.7473
96	92 16	9.7980	139	1 93 21	11.7898
97	94 09	9.8489	140	1 96 00	11.8322
98	96 04	9.8995			
99	98 01	9.9499	141	1 98 81	11.8743
100	1 00 00	10.0000	142	2 01 64	11.9164
			143	2 04 49	11.9583
101	1 02 01	10.0499	144	2 07 36	12.0000
102	1 04 04	10.0995	145	2 10 25	12.0416
103	1 06 09	10.1489			
104	1 08 16	10.1982	146	2 13 16	12.0830
105	1 10 25	10.2470	147	2 16 09	12.1244
			148	2 19 04	12.1655
106	1 12 36	10.2956	149	2 22 01	12.0266
107	1 14 49	10.3441	150	2 25 00	12.2474
108	1 16 64	10.3923			
109	1 18 81	10.4403	151	2 28 01	12.2882
110	1 21 00	10.4881	152	2 31 04	12.3288
			153	2 34 09	12.3693
111	1 23 21	10.5357	154	2 37 16	12.4097
112	1 25 44	10.5830	155	2 40 25	12.4499
113	1 27 69	10.6301			
114	1 29 96	10.6771	156	2 43 36	12.4900
115	1 32 25	10.7238	157	2 46 49	12.5300
			158	2 49 64	12.5698
116	1 34 56	10.7703	159	2 52 81	12.6095
117	1 36 89	10.8167	160	2 56 00	12.6491
118	1 39 24	10.8628			
119	1 41 61	10.9087	161	2 59 21	12.6886
120	1 44 00	10.9545	162	2 62 44	12.7279
			163	2 65 69	12.7671
121	1 46 41	11.0000	164	2 68 96	12.8062
122	1 48 84	11.0454	165	2 72 25	12.8452
123	1 51 29	11.0905			
124	1 58 76	11.1355	166	2 75 56	12.8841
125	1 56 25	11.1803	167	2 78 89	12.9228
			168	2 82 24	12.9615
126	1 58 76	11.2250	169	2 85 61	13.0000
127	1 61 29	11.2694	170	2 89 00	13.0384
128	1 63 84	11.3137			

N	N²	√N	N	N²	√N
171	2 92 41	13.0767	214	4 57 96	14.6287
172	2 95 84	13.1149	215	4 62 25	14.6629
173	2 99 29	13.1529			
174	3 02 76	13.1909	216	4 66 56	14.6969
175	3 06 25	13.2288	217	4 70 89	14.7309
			218	4 75 24	14.7648
176	3 09 76	13.2665	219	4 79 61	14.7986
177	3 13 29	13.3041	220	4 84 00	14.8324
178	3 16 84	13.3417			
179	3 20 41	13.3791	221	4 88 41	14.8661
180	3 24 00	13.4164	222	4 92 84	14.8997
			223	4 97 29	14.9332
181	3 27 61	13.4536	224	5 01 76	14.9666
182	3 31 24	13.4907	225	5 06 25	15.0000
183	3 34 89	13.5277			
184	3 38 56	13.5647	226	5 10 76	15.0333
185	3 42 25	13.6015	227	5 15 29	15.0665
			228	5 19 84	15.0997
186	3 45 96	13.6382	229	5 24 41	15.1327
187	3 49 69	13.6748	230	5 29 00	15.1658
188	3 53 44	13.7113			
189	3 57 21	13.7477	231	5 33 61	15.1987
190	3 61 00	13.7840	232	5 38 24	15.2315
			233	5 42 89	15.2643
191	3 64 81	13.8203	234	5 47 56	15.2971
192	3 68 64	13.8564	235	5 52 25	15.3297
193	3 72 49	13.8924			
194	3 76 36	13.9284	236	5 56 96	15.3623
195	3 80 25	13.9642	237	5 61 69	15.3948
			238	5 66 44	15.4272
196	3 84 16	14.0000	239	5 71 21	15.4596
197	3 88 09	14.0357	240	5 76 00	15.4919
198	3 92 04	14.0712			
199	3 96 01	14.1067	241	5 80 81	15.5242
200	4 00 00	14.1421	242	5 85 64	15.5563
			243	5 90 49	15.5885
201	4 04 01	14.1774	244	5 95 36	15.6205
202	4 08 04	14.2127	245	6 00 25	15.6525
203	4 12 09	14.2478			
204	4 16 16	14.2829	246	6 05 16	15.6844
205	4 20 25	14.3178	247	6 10 09	15.7162
			248	6 15 04	15.7480
206	4 24 36	14.3527	249	6 20 01	15.7797
207	4 28 49	14.3875	250	6 25 00	15.8114
208	4 32 64	14.4222			
209	4 36 81	14.4568	251	6 30 01	15.8430
210	4 41 00	14.4914	252	6 35 04	15.8745
			253	6 40 09	15.9060
211	4 45 21	14.5258	254	6 45 16	15.9374
212	4 49 44	14.5602	255	6 50 25	15.9687
213	4 53 69	14.5945			

N	N²	√N	N	N²	√N
256	6 55 36	16.0000	299	8 94 01	17.2916
257	6 60 49	16.0321	300	9 00 00	17.3205
258	6 65 64	16.0624			
259	6 70 81	16.0935	301	9 06 01	17.3494
260	6 76 00	16.1245	302	9 12 04	17.3781
			303	9 18 09	17.4069
261	6 81 21	16.1555	304	9 24 16	17.4356
262	6 86 44	16.1864	305	9 30 25	17.4642
263	6 91 69	16.2173			
264	6 96 96	16.2481	306	9 36 36	17.4929
265	7 02 25	16.2788	307	9 42 49	17.5214
			308	9 48 64	17.5499
266	7 07 56	16.3095	309	9 54 81	17.5784
267	7 12 89	16.3401	310	9 61 00	17.6068
268	7 18 24	16.3707			
269	7 23 61	16.4012	311	9 67 21	17.6352
270	7 29 00	16.4317	312	9 73 44	17.6635
			313	9 79 69	17.6918
271	7 34 41	16.4621	314	9 85 96	17.7200
272	7 39 84	16.4924	315	9 92 25	17.7482
273	7 45 29	16.5227			
274	7 50 76	16.5529	316	9 98 56	17.7764
275	7 56 25	16.5831	317	10 04 89	17.8045
			318	10 11 24	17.8326
276	7 61 76	16.6132	319	10 17 61	17.8606
277	7 67 29	16.6433	320	10 24 00	17.8885
278	7 72 84	16.6733			
279	7 78 41	16.7033	321	10 30 41	17.9165
280	7 84 00	16.7332	322	10 36 84	17.9444
			323	10 43 29	17.9722
281	7 89 61	16.7631	324	10 49 76	18.0000
282	7 95 24	16.7929	325	10 56 25	18.0278
283	8 00 89	16.8226			
284	8 06 56	16.8523	326	10 62 76	18.0555
285	8 12 25	16.8819	327	10 69 29	18.0831
			328	10 75 84	18.1108
286	8 17 96	16.9115	329	10 82 41	18.1384
287	8 23 69	16.9411	330	10 89 00	18.1659
288	8 28 44	16.9706			
289	8 35 21	17.0000	331	10 95 61	18.1934
290	8 41 00	17.0294	332	11 02 24	18.2209
			333	11 08 89	18.2483
291	8 46 81	17.0587	334	11 15 56	18.2757
292	8 52 64	17.0880	335	11 22 25	18.3030
293	8 58 49	17.1172			
294	8 64 36	17.1464	336	11 28 96	18.3303
295	8 70 25	17.1756	337	11 35 69	18.3576
			338	11 42 44	18.3848
296	8 76 16	17.2047	339	11 49 21	18.4120
297	8 82 09	17.2337	340	11 56 00	18.4391
298	8 88 04	17.2627			

N	N^2	$\sqrt{N}$	N	N^2	$\sqrt{N}$
341	11 62 81	18.4662	384	14 74 56	19.5959
342	11 69 64	18.4932	385	14 82 25	19.6214
343	11 76 49	18.5203			
344	11 83 36	18.5472	386	14 89 96	19.6469
345	11 90 25	18.5742	387	14 97 69	19.6723
			388	15 05 44	19.6977
346	11 97 16	18.6011	389	15 13 21	19.7231
347	12 04 09	18.6279	390	15 21 00	19.7484
348	12 11 04	18.6548			
349	12 18 01	18.6815	391	15 28 81	19.7737
350	12 25 00	18.7083	392	15 36 64	19.7990
			393	15 44 49	19.8242
351	12 32 01	18.7350	394	15 52 36	19.8494
352	12 39 04	18.7617	395	15 60 25	19.8746
353	12 46 09	18.7883			
354	12 53 16	18.8149	396	15 68 16	19.8997
355	12 60 25	18.8414	397	15 76 09	19.9249
			398	15 84 04	19.9499
356	12 67 36	18.8680	399	15 92 01	19.9750
357	12 74 49	18.8944	400	16 00 00	20.0000
358	12 81 64	18.9209			
359	12 88 81	18.9473	401	16 08 01	20.0250
360	12 96 00	18.9737	402	16 16 04	20.0499
			403	16 24 09	20.0749
361	13 03 21	19.0000	404	16 32 16	20.0998
362	13 01 44	19.0263	405	16 40 25	20.1246
363	13 17 69	19.0526			
364	13 24 96	19.0788	406	16 48 36	20.1494
365	13 32 25	19.1050	407	16 56 49	20.1742
			408	16 64 64	20.1990
366	13 39 56	19.1311	409	16 72 81	20.2237
367	13 46 89	19.1572	410	16 81 00	20.2485
368	13 54 24	19.1833			
369	13 61 61	19.2094	411	16 89 21	20.2731
370	13 69 00	19.2354	412	16 97 44	20.2978
			413	17 05 69	20.3224
371	13 76 41	19.2614	414	17 13 96	20.3470
372	13 83 84	19.2873	415	17 22 25	20.3715
373	13 91 29	19.3132			
374	13 98 76	19.3391	416	17 30 56	20.3961
375	14 06 25	19.3649	417	17 38 89	20.4206
			418	17 47 24	20.4450
376	14 13 76	19.3907	419	17 55 61	20.4695
377	14 21 29	19.4165	420	17 64 00	20.4939
378	14 28 84	19.4422			
379	14 36 41	19.4679	421	17 72 41	20.5183
380	14 44 00	19.4936	422	17 80 84	20.5426
			423	17 89 29	20.5670
381	14 51 61	19.5192	424	17 97 76	20.5913
382	14 59 24	19.5449	425	18 06 25	20.6155
383	14 66 89	19.5704			

N	N²	√N	N	N²	√N
426	18 14 76	20.6398	469	21 99 61	21.6564
427	18 23 29	20.6640	470	22 09 00	21.6795
428	18 31 84	20.6882			
429	18 40 41	20.7123	471	22 18 41	21.7025
430	18 49 00	20.7364	472	22 27 84	21.7256
			473	22 37 29	21.7486
431	18 57 61	20.7605	474	22 46 76	21.7715
432	18 66 24	20.7846	475	22 56 25	21.7945
433	18 74 89	20.8087			
434	18 83 56	20.8327	476	22 65 76	21.8174
435	18 92 25	20.8567	477	22 75 29	21.8403
			478	22 84 84	21.8632
436	19 00 06	20.8806	479	22 94 41	21.8861
437	19 09 69	20.9045	480	23 04 00	21.9089
438	19 18 44	20.9284			
439	19 27 21	20.9523	481	23 13 61	21.9317
440	19 36 00	20.9762	482	23 23 24	21.9545
			483	23 32 89	21.9773
441	19 44 81	21.0000	484	23 42 56	22.0000
442	19 53 64	21.0238	485	23 52 25	22.0227
443	19 62 49	21.0476			
444	19 71 36	21.0713	486	23 61 96	22.0454
445	19 80 25	21.0950	487	23 71 69	22.0681
			488	23 81 44	22.0907
446	19 89 16	21.1187	489	23 91 21	22.1133
447	19 98 09	21.1424	490	24 01 00	22.1359
448	20 07 04	21.1660			
449	20 16 01	21.1896	491	24 10 81	22.1585
450	20 25 00	21.2132	492	24 20 64	22.1811
			493	24 30 49	22.2036
451	20 34 01	21.2368	494	24 40 36	22.2261
452	20 43 04	21.2603	495	24 50 25	22.2486
453	20 52 09	21.2838			
454	20 61 16	21.3073	496	24 60 16	22.2711
455	20 70 25	21.3307	497	24 70 09	22.2935
			498	24 80 04	22.3159
456	20 79 36	21.3542	499	24 90 01	22.3383
457	20 88 49	21.3776	500	25 00 00	22.3607
458	20 97 64	21.4009			
459	21 06 81	21.4243	501	25 10 01	22.3830
460	21 16 00	21.4476	502	25 20 04	22.4054
			503	25 30 09	22.4277
461	21 25 21	21.4709	504	25 40 16	22.4499
462	21 34 44	21.4942	505	25 50 25	22.4722
463	21 43 69	21.5174			
464	21 52 96	21.5407	506	25 60 36	22.4944
465	21 62 25	21.5639	507	25 70 49	22.5167
			508	25 80 64	22.5389
466	21 71 56	21.5870	509	25 90 81	22.5610
467	21 80 89	21.6102	510	26 01 00	22.5832
468	21 90 24	21.6333			

N	N²	√N	N	N²	√N
511	26 11 21	22.6053	554	30 69 16	23.5372
512	26 21 44	22.6274	555	30 80 25	23.5584
513	26 31 69	22.6495			
514	26 41 96	22.6716	556	30 91 36	23.5797
515	26 52 25	22.6936	557	31 02 49	23.6008
			558	31 13 64	23.6220
516	26 62 56	22.7156	559	31 24 81	23.6432
517	26 72 89	22.7376	560	31 36 00	23.6643
518	26 83 24	22.7596			
519	26 93 61	22.7816	561	31 47 21	23.6854
520	27 04 00	22.8035	562	31 58 44	23.7065
			563	31 69 69	23.7276
521	27 14 41	22.8254	564	31 80 96	23.7487
522	27 24 84	22.8473	565	31 92 25	23.7697
523	27 35 29	22.8692			
524	27 45 76	22.8910	566	32 03 56	23.7908
525	27 56 25	22.9129	567	32 14 89	23.9118
			568	32 26 24	23.8328
526	27 66 76	22.9347	569	32 37 61	23.8537
527	27 77 29	22.9565	570	32 49 00	23.8747
528	27 87 84	22.9783			
529	27 98 41	23.0000	571	32 60 41	23.8956
530	28 09 00	23.0217	572	32 71 84	23.916
			573	32 83 29	23.9374
531	28 19 61	23.0434	574	32 94 76	23.9583
532	28 30 24	23.0651	575	33 06 25	23.9792
533	28 40 89	23.0868			
534	28 51 56	23.1084	576	33 17 76	24.0000
535	28 62 25	23.1301	577	33 29 29	24.0208
			578	33 40 84	24.0416
536	28 72 96	23.1517	579	33 52 41	24.0624
537	28 83 69	23.1733	580	33 64 00	24.0832
538	28 94 44	23.1948			
539	29 05 21	23.2164	581	33 75 61	24.1039
540	29 16 00	23.2379	582	33 87 24	24.1247
			583	33 98 89	24.1454
541	29 26 81	23.2594	584	34 10 56	24.1661
542	29 37 64	23.2809	585	34 22 25	24.1868
543	29 48 49	23.3024			
544	29 59 36	23.3238	586	34 33 96	24.2074
545	29 70 25	23.3452	587	34 45 69	24.2281
			588	34 57 44	24.2487
546	29 81 16	23.3666	589	34 69 21	24.2693
547	29 92 09	23.3880	590	34 81 00	24.2899
548	30 03 04	23.4094			
549	30 14 01	23.4307	591	34 92 81	24.3105
550	30 25 00	23.4521	592	35 04 64	24.3311
			593	35 16 49	24.3516
551	30 36 01	23.4734	594	35 28 36	24.3721
552	30 47 04	23.4947	595	35 40 25	24.3926
553	30 58 09	23.5160			

N	N^2	$\sqrt{N}$	N	N^2	$\sqrt{N}$
596	35 52 16	24.4131	639	40 83 21	25.2784
597	35 64 09	24.4336	640	40 96 00	25.2982
598	35 76 04	24.4540			
599	35 88 01	24.4745	641	41 08 81	25.3180
600	36 00 00	24.4949	642	41 21 64	25.3377
			643	41 34 49	25.3574
601	36 12 01	24.5153	644	41 47 36	25.3772
602	36 24 04	24.5357	645	41 60 25	25.3969
603	36 36 09	24.5561			
604	36 48 16	24.5764	646	41 73 16	25.4165
605	36 60 25	24.5967	647	41 86 09	25.4362
			648	41 99 04	25.4558
606	36 72 36	24.6171	649	42 12 01	25.4775
607	36 84 49	24.6374	650	42 25 00	25.4951
608	36 96 64	24.6577			
609	37 08 81	24.6779	651	42 38 01	25.5147
610	37 21 00	24.6982	652	42 51 04	25.5343
			653	42 64 09	25.5539
611	37 33 21	24.7184	654	42 77 16	25.5734
612	37 45 44	24.7386	655	42 90 25	25.5930
613	37 57 69	24.7588			
614	37 69 96	24.7790	656	43 03 36	25.6125
615	37 82 25	24.7992	657	43 16 49	25.6320
			658	43 29 64	25.6515
616	37 94 56	24.8193	659	43 42 81	25.6710
617	38 06 89	24.8395	660	43 56 00	25.6905
618	38 19 24	24.8596			
619	38 31 61	24.8797	661	43 69 21	25.7099
620	38 44 00	24.8998	662	43 82 44	25.7294
			663	43 95 69	25.7488
621	38 56 41	24.9199	664	44 08 96	25.7682
622	38 68 84	24.9399	665	44 22 25	25.7876
623	38 81 29	24.9600			
624	38 93 76	24.9800	666	44 35 56	25.8070
625	39 06 25	25.0000	667	44 48 89	25.8263
			668	44 62 24	25.8457
626	39 18 76	25.0200	669	44 75 61	25.8650
627	39 31 29	25.0400	670	44 89 00	25.8844
628	39 43 84	25.0599			
629	39 56 41	25.0799	671	45 02 41	25.9037
630	39 69 00	25.0998	672	45 15 84	25.9230
			673	45 29 29	25.9422
631	39 81 61	25.1197	674	45 42 76	25.9615
632	39 94 24	25.1396	675	45 56 25	25.9808
633	40 06 89	25.1595			
634	40 19 56	25.1794	676	45 69 76	26.0000
635	40 32 25	25.1992	677	45 83 29	26.0192
			678	45 96 84	26.0384
636	40 44 96	25.2190	679	46 10 41	26.0576
637	40 57 69	25.2389	680	46 24 00	26.0768
638	40 70 44	25.2587			

N	N²	√N	N	N²	√N
681	46 37 61	26.0960	724	52 41 76	26.9072
682	46 51 24	26.1151	725	52 56 25	26.9258
683	46 64 89	26.1343			
684	46 78 56	26.1534	726	52 70 76	26.9444
685	46 92 25	26.1725	727	52 85 29	26.9629
			728	52 99 84	26.9815
686	47 05 96	26.1916	729	53 14 41	27.0000
687	47 19 69	26.2107	730	53 29 00	27.0185
688	47 33 44	26.2298			
689	47 47 21	26.2488	731	53 43 61	27.0370
690	47 61 00	26.2679	732	53 58 24	27.0555
			733	53 72 89	27.0740
691	47 74 81	26.2869	734	53 87 56	27.0924
692	47 88 64	26.3059	735	54 02 25	27.1109
693	48 02 49	26.3249			
694	48 16 36	26.3439	736	54 16 96	27.1293
695	48 30 25	26.3629	737	54 31 69	27.1477
			738	54 46 44	27.1662
696	48 44 16	26.3818	739	54 61 21	27.1846
697	48 58 09	26.4008	740	54 76 00	27.2029
698	48 72 04	26.4197			
699	48 86 01	26.4386	741	54 90 81	27.2213
700	49 00 00	26.4575	742	55 05 64	27.2397
			743	55 20 49	27.2580
701	49 14 01	26.4764	744	55 35 36	27.2764
702	49 28 04	26.4953	745	55 50 25	27.2947
703	49 42 09	26.5141			
704	49 56 16	26.5330	746	55 65 16	27.3130
705	49 70 25	26.5518	747	55 80 09	27.3313
			748	55 95 04	27.3496
706	49 84 36	26.5707	749	56 10 01	27.3679
707	49 98 49	26.5895	750	56 25 00	27.3861
708	50 12 64	26.6083			
709	50 26 81	26.6271	751	56 40 01	27.4044
710	50 41 00	26.6458	752	56 55 04	27.4226
			753	56 70 09	27.4408
711	50 55 21	26.6646	754	56 85 16	27.4591
712	50 69 44	26.6833	755	57 00 25	27.4773
713	50 83 69	26.7021			
714	50 97 96	26.7208	756	57 15 36	27.4955
715	51 12 25	26.7395	757	57 30 49	27.5136
			758	57 45 64	27.5318
716	51 26 56	26.7582	759	57 60 81	27.5500
717	51 40 89	26.7769	760	57 76 00	27.5681
718	51 55 24	26.7955			
719	51 69 61	26.8142	761	57 91 21	27.5862
720	51 84 00	26.8328	762	58 06 44	27.6043
			763	58 21 69	27.6225
721	51 98 41	26.8514	764	58 36 96	27.6405
722	52 12 84	26.8701	765	58 52 25	27.6586
723	52 27 29	26.8887			

N	N²	√N	N	N²	√N
766	58 67 56	27.6767	809	65 44 81	28.4429
767	58 82 89	27.6948	810	65 61 00	28.4605
768	58 98 24	27.7128			
769	59 13 61	27.7308	811	65 77 21	28.4781
770	59 29 00	27.7489	812	65 93 44	28.4956
			813	66 09 69	28.5132
771	59 44 41	27.7669	814	66 25 96	28.5307
772	59 59 84	27.7849	815	66 42 25	28.5482
773	59 75 29	27.8029			
774	59 90 76	27.8209	816	66 58 56	28.5657
775	60 06 25	27.8388	817	66 74 89	28.5832
			818	66 91 24	28.6007
776	60 21 76	27.8568	819	67 07 61	28.6182
777	60 37 29	27.8747	820	67 24 00	28.6356
778	60 52 84	27.8927			
779	60 68 41	27.9106	821	67 40 41	28.6531
780	60 84 00	27.9285	822	67 56 84	28.6705
			823	67 73 29	28.6880
781	60 99 61	27.9464	824	67 89 76	28.7054
782	61 15 24	27.9643	825	68 06 25	28.7228
783	61 30 89	27.9821			
784	61 46 56	28.0000	826	68 22 67	28.7402
785	61 62 25	28.0179	827	68 39 29	28.7576
			828	68 55 84	28.7750
786	61 77 96	28.0357	829	68 72 41	28.7924
787	61 93 69	28.0535	830	68 89 00	28.8097
788	62 09 44	28.0713			
789	62 25 21	28.0891	831	69 05 61	28.8271
790	62 41 00	28.1069	832	69 22 24	28.8444
			833	69 38 89	28.8617
791	62 56 81	28.1247	834	69 55 56	28.8791
792	62 72 64	28.1425	835	69 72 25	28.8964
793	62 88 49	28.1603			
794	63 04 36	28.1780	836	69 88 96	28.9137
795	63 20 25	28.1957	837	70 05 69	28.9310
			838	70 22 44	28.9482
796	63 36 16	28.2135	839	70 39 21	28.9655
797	63 52 09	28.2312	840	70 56 00	28.9828
798	63 68 04	28.2489			
799	63 84 01	28.2666	841	70 72 81	29.0000
800	64 00 00	28.2843	842	70 89 64	29.0172
			843	71 06 49	29.0345
801	64 16 01	28.3019	844	71 23 36	29.0517
802	64 32 04	28.3196	845	71 40 25	29.0689
803	64 48 09	28.3373			
804	64 64 16	28.3549	846	71 57 16	29.0861
805	64 80 25	28.3725	847	71 74 09	29.1033
			848	71 91 04	29.1204
806	64 96 36	28.3901	849	72 08 01	29.1376
807	65 12 49	28.4077	850	72 25 00	29.1548
808	65 28 64	28.4253			

N	N²	√N	N	N²	√N
851	72 42 01	29.1719	894	79 92 36	29.8998
852	72 59 04	29.1890	895	80 10 25	29.9166
853	72 76 09	29.2062			
854	72 93 16	29.2233	896	80 28 16	29.9333
855	73 10 25	29.2404	897	80 46 09	29.9500
			898	80 64 04	29.9666
856	73 27 36	29.2575	899	80 82 01	29.9833
857	73 44 49	29.2746	900	81 00 00	30.0000
858	73 61 64	29.2916			
859	73 78 81	29.3087	901	81 18 01	30.0167
860	73 96 00	29.3258	902	81 36 04	30.0333
			903	81 54 09	30.0500
861	74 13 21	29.3428	904	81 72 16	30.0666
862	74 30 44	29.3598	905	81 90 25	30.0832
863	74 47 69	29.3769			
864	74 64 96	29.3939	906	82 08 36	30.0998
865	74 82 25	29.4109	907	82 26 49	30.1164
			908	82 44 64	30.1330
866	74 99 56	29.4279	909	82 62 81	30.1496
867	75 16 89	29.4449	910	82 81 00	30.1662
868	75 34 24	29.4618			
869	75 51 61	29.4788	911	82 99 21	30.1828
870	75 69 00	29.4958	912	83 17 44	30.1993
			913	83 35 69	30.2159
871	75 86 41	29.5127	914	83 53 96	30.2324
872	76 03 84	29.5296	915	83 72 25	30.2490
873	76 21 29	29.5466			
874	76 38 76	29.5635	916	83 90 56	30.2655
875	76 56 25	29.5804	917	84 08 89	30.2820
			918	84 27 24	30.2985
876	76 73 76	29.5973	919	84 45 61	30.3150
877	76 91 29	29.6142	920	84 64 00	30.3315
878	77 08 84	29.6311			
879	77 26 41	29.6479	921	84 82 41	30.3480
880	77 44 00	29.6648	922	85 00 84	30.3645
			923	85 19 29	30.3809
881	77 61 61	29.6816	924	85 37 76	30.3974
882	77 79 24	29.6985	925	85 56 25	30.4138
883	77 96 89	29.7153			
884	78 14 56	29.7321	926	85 74 76	30.4302
885	78 32 25	29.7489	927	85 93 29	30.4467
			928	86 11 84	30.4631
886	78 49 96	29.7658	929	86 30 41	30.4795
887	78 67 69	29.7825	930	86 49 00	30.4959
888	78 85 44	29.7993			
889	79 03 21	29.8161	931	86 67 61	30.5123
890	79 21 00	29.8329	932	86 86 24	30.5287
			933	87 04 89	30.5450
891	79 38 81	29.8496	934	87 23 56	30.5614
892	79 56 64	29.8664	935	87 42 25	30.5778
893	79 74 49	29.8831			

N	N^2	$\sqrt{N}$	N	N^2	$\sqrt{N}$
936	87 50 96	30.5941	969	93 89 61	31.1288
937	87 79 69	30.6105	970	94 09 00	31.1448
938	87 98 44	30.6268			
939	88 17 21	30.6431	971	94 28 41	31.1609
940	88 36 00	30.6594	972	94 47 84	31.1769
			973	94 67 29	31.1929
941	88 54 81	30.6757	974	94 86 76	31.2090
942	88 73 64	30.6920	975	95 06 25	31.2250
943	88 92 49	30.7083			
944	89 11 36	30.7246	976	95 25 76	31.2410
945	89 30 25	30.7409	977	95 45 29	31.2570
			978	95 64 84	31.2730
946	89 49 16	30.7571	979	95 84 41	31.2890
947	89 68 09	30.7734	980	96 04 00	31.3050
948	89 87 04	30.7896			
949	90 06 01	30.8058	981	96 23 61	31.3209
950	90 25 00	30.8221	982	96 43 24	31.3369
			983	96 62 89	31.3528
951	90 44 01	30.8383	984	96 82 56	31.3688
952	90 63 04	30.8545	985	97 02 25	31.3847
953	90 82 09	30.8707			
954	91 01 16	30.8869	986	97 21 96	31.4006
955	91 20 25	30.9031	987	97 41 69	31.4166
			988	97 61 44	31.4325
956	91 39 36	30.9192	989	97 81 21	31.4484
957	91 58 49	30.9354	990	98 01 00	31.4643
958	91 77 64	30.9516			
959	91 96 81	30.9677	991	98 20 81	31.4802
960	92 16 00	30.9839	992	98 40 64	31.4960
			993	98 60 49	31.5119
961	92 35 21	31.0000	994	98 80 36	31.5278
962	92 54 44	31.0161	995	99 00 25	31.5436
963	92 73 69	31.0322			
964	92 92 96	31.0483	996	99 20 16	31.5595
965	93 12 25	31.0644	997	99 40 09	31.5753
			998	99 60 04	31.5911
966	93 31 56	31.0805	999	99 80 01	31.6070
967	93 50 89	31.0966	1000	1 00 00 00	31.6228
968	93 70 24	31.1127			

TABLE P. Four-Place Common Logarithms of Numbers

No. X	Log X										Proportional Parts								
	0	1	2	3	4	5	6	7	8	9	1	2	3	4	5	6	7	8	9
100	0000	0004	0009	0013	0017	0022	0026	0030	0035	0039	0	1	1	2	2	3	3	3	4
101	0043	0048	0052	0056	0060	0065	0069	0073	0077	0082	0	1	1	2	2	3	3	3	4
102	0086	0090	0095	0099	0103	0107	0111	0116	0120	0124	0	1	1	2	2	3	3	3	4
103	0128	0133	0137	0141	0145	0149	0154	0158	0162	0166	0	1	1	2	2	3	3	3	4
104	0170	0175	0179	0183	0187	0191	0195	0199	0204	0208	0	1	1	2	2	2	3	3	4
105	0212	0216	0220	0224	0228	0233	0237	0241	0245	0249	0	1	1	2	2	2	3	3	4
106	0253	0257	0261	0265	0269	0273	0278	0282	0286	0290	0	1	1	2	2	2	3	3	4
107	0294	0298	0302	0306	0310	0314	0318	0322	0326	0330	0	1	1	2	2	2	3	3	4
108	0334	0338	0342	0346	0350	0354	0358	0362	0366	0370	0	1	1	2	2	2	3	3	4
109	0374	0378	0382	0386	0390	0394	0398	0402	0406	0410	0	1	1	2	2	2	3	3	4
10	0000	0043	0086	0128	0170	0212	0253	0294	0334	0374	4	8	12	17	21	25	29	33	37
11	0414	0453	0492	0531	0569	0607	0645	0682	0719	0755	4	8	11	15	19	23	26	30	34
12	0792	0828	0864	0899	0934	0969	1004	1038	1072	1106	3	7	10	14	17	21	24	28	31
13	1139	1173	1206	1239	1271	1303	1335	1367	1399	1430	3	6	10	13	16	19	23	26	29
14	1461	1492	1523	1553	1584	1614	1644	1673	1703	1732	3	6	9	12	15	18	21	24	27
15	1761	1790	1818	1847	1875	1903	1931	1959	1987	2014	3	6	8	11	14	17	20	22	25
16	2041	2068	2095	2122	2148	2175	2201	2227	2253	2279	3	5	8	11	13	16	18	21	24
17	2304	2330	2355	2380	2405	2430	2455	2480	2504	2529	2	5	7	10	12	15	17	20	22
18	2553	2577	2601	2625	2648	2672	2695	2718	2742	2765	2	5	7	9	12	14	16	19	21
19	2788	2810	2833	2856	2878	2900	2923	2945	2967	2989	2	4	7	9	11	13	16	18	20
20	3010	3032	3054	3075	3096	3118	3139	3160	3181	3201	2	4	6	8	11	13	15	17	19
21	3222	3243	3263	3284	3304	3324	3345	3365	3385	3404	2	4	6	8	10	12	14	16	18
22	3424	3444	3464	3483	3502	3522	3541	3560	3579	3598	2	4	6	8	10	12	14	15	17
23	3617	3636	3655	3674	3692	3711	3729	3747	3766	3784	2	4	6	7	9	11	13	15	17
24	3802	3820	3838	3856	3874	3892	3909	3927	3945	3962	2	4	5	7	9	11	12	14	16

Logarithms

X	0	1	2	3	4	5	6	7	8	9	1	2	3	4	5	6	7	8	9
25	3979	3997	4014	4031	4048	4065	4082	4099	4116	4133	2	3	5	7	9	10	12	14	15
26	4150	4166	4183	4200	4216	4232	4249	4265	4281	4298	2	3	5	7	8	10	12	13	15
27	4314	4330	4346	4362	4378	4393	4409	4425	4440	4456	2	3	5	6	8	9	11	13	14
28	4472	4487	4502	4518	4533	4548	4564	4579	4594	4609	2	3	5	6	8	9	11	12	14
29	4624	4639	4654	4669	4683	4698	4713	4728	4742	4757	1	3	4	6	7	9	10	12	13
30	4771	4786	4800	4814	4829	4843	4857	4871	4886	4900	1	3	4	6	7	9	10	11	13
31	4914	4928	4942	4955	4969	4983	4997	5011	5024	5038	1	3	4	6	7	8	10	11	12
32	5051	5065	5079	5092	5105	5119	5132	5145	5159	5172	1	3	4	5	7	8	9	11	12
33	5185	5198	5211	5224	5237	5250	5263	5276	5289	5302	1	3	4	5	6	8	9	10	12
34	5315	5328	5340	5353	5366	5378	5391	5403	5416	5428	1	3	4	5	6	8	9	10	11
35	5441	5453	5465	5478	5490	5502	5514	5527	5539	5551	1	2	4	5	6	7	9	10	11
36	5563	5575	5587	5599	5611	5623	5635	5647	5658	5670	1	2	4	5	6	7	8	10	11
37	5682	5694	5705	5717	5729	5740	5752	5763	5775	5786	1	2	3	5	6	7	8	9	10
38	5798	5809	5821	5832	5843	5855	5866	5877	5888	5899	1	2	3	5	6	7	8	9	10
39	5911	5922	5933	5944	5955	5966	5977	5988	5999	6010	1	2	3	4	5	7	8	9	10
40	6021	6031	6042	6053	6064	6075	6085	6096	6107	6117	1	2	3	4	5	6	8	9	10
41	6128	6138	6149	6160	6170	6180	6191	6201	6212	6222	1	2	3	4	5	6	7	8	9
42	6232	6243	6253	6263	6274	6284	6294	6304	6314	6325	1	2	3	4	5	6	7	8	9
43	6335	6345	6355	6365	6375	6385	6395	6405	6415	6425	1	2	3	4	5	6	7	8	9
44	6435	6444	6454	6464	6474	6484	6493	6503	6513	6522	1	2	3	4	5	6	7	8	9
45	6532	6542	6551	6561	6571	6580	6590	6599	6609	6618	1	2	3	4	5	6	7	8	9
46	6628	6637	6646	6656	6665	6675	6684	6693	6702	6712	1	2	3	4	5	6	7	7	8
47	6721	6730	6739	6749	6758	6767	6776	6785	6794	6803	1	2	3	4	5	5	6	7	8
48	6812	6821	6830	6839	6848	6857	6866	6875	6884	6893	1	2	3	4	5	5	6	7	8
49	6902	6911	6920	6928	6937	6946	6955	6964	6972	6981	1	2	3	4	4	5	6	7	8

No. X	Log X										Proportional Parts								
	0	1	2	3	4	5	6	7	8	9	1	2	3	4	5	6	7	8	9
50	6990	6998	7007	7016	7024	7033	7042	7050	7059	7067	1	2	3	3	4	5	6	7	8
51	7076	7084	7093	7101	7110	7118	7126	7135	7143	7152	1	2	3	3	4	5	6	7	8
52	7160	7168	7177	7185	7193	7202	7210	7218	7226	7235	1	2	2	3	4	5	6	7	8
53	7243	7251	7259	7267	7275	7284	7292	7300	7308	7316	1	2	2	3	4	5	6	6	7
54	7324	7332	7340	7348	7356	7364	7372	7380	7388	7396	1	2	2	3	4	5	6	6	7
55	7404	7412	7419	7427	7435	7443	7451	7459	7466	7474	1	2	2	3	4	5	5	6	7
56	7482	7490	7497	7505	7513	7520	7528	7536	7543	7551	1	2	2	3	4	5	5	6	7
57	7559	7566	7574	7582	7589	7597	7604	7612	7619	7627	1	2	2	3	4	5	5	6	7
58	7634	7642	7649	7657	7664	7672	7679	7686	7694	7701	1	1	2	3	4	4	4	6	7
59	7709	7716	7723	7731	7738	7745	7752	7760	7767	7774	1	1	2	3	4	4	5	6	7
60	7782	7789	7796	7803	7810	7818	7825	7832	7839	7846	1	1	2	3	4	4	5	6	6
61	7853	7860	7868	7875	7882	7889	7896	7903	7910	7917	1	1	2	3	4	4	5	6	6
62	7924	7931	7938	7945	7952	7959	7966	7973	7980	7987	1	1	2	3	3	4	5	6	6
63	7993	8000	8007	8014	8021	8028	8035	8041	8048	8055	1	1	2	3	3	4	5	5	6
64	8062	8069	8075	8082	8089	8096	8102	8109	8116	8122	1	1	2	3	3	4	5	5	6
65	8129	8136	8142	8149	8156	8162	8169	8176	8182	8189	1	1	2	3	3	4	5	5	6
66	8195	8202	8209	8215	8222	8228	8235	8241	8248	8254	1	1	2	3	3	4	5	5	6
67	8261	8267	8274	8280	8287	8293	8299	8306	8312	8319	1	1	2	3	3	4	5	5	6
68	8325	8331	8338	8344	8351	8357	8363	8370	8376	8382	1	1	2	3	3	4	4	5	6
69	8388	8395	8401	8407	8414	8420	8426	8432	8439	8445	1	1	2	2	3	4	4	5	6
70	8451	8457	8463	8470	8476	8482	8488	8494	8500	8506	1	1	2	2	3	4	4	5	6
71	8513	8519	8525	8531	8537	8543	8549	8555	8561	8567	1	1	2	2	3	4	4	5	5
72	8573	8579	8585	8591	8597	8603	8609	8615	8621	8627	1	1	2	2	3	4	4	5	5
73	8633	8639	8645	8651	8657	8663	8669	8675	8681	8686	1	1	2	3	3	4	4	5	5

X	0	1	2	3	4	5	6	7	8	9	1	2	3	4	5	6	7	8	9
74	8692	8698	8704	8710	8716	8722	8727	8733	8739	8745	1	1	2	2	3	4	4	5	5
75	8751	8756	8762	8768	8774	8779	8785	8791	8797	8802	1	1	2	2	3	3	4	5	5
76	8808	8814	8820	8825	8831	8837	8842	8848	8854	8859	1	1	2	2	3	3	4	5	5
77	8865	8871	8876	8882	8887	8893	8899	8904	8910	8915	1	1	2	2	3	3	4	4	5
78	8921	8927	8932	8938	8943	8949	8954	8960	8965	8971	1	1	2	2	3	3	4	4	5
79	8976	8982	8987	8993	8998	9004	9009	9015	9020	9025	1	1	2	2	3	3	4	4	5
80	9031	9036	9042	9047	9053	9058	9063	9069	9074	9079	1	1	2	2	3	3	4	4	5
81	9085	9090	9096	9101	9106	9112	9117	9122	9128	9133	1	1	2	2	3	3	4	4	5
82	9138	9143	9149	9154	9159	9165	9170	9175	9180	9186	1	1	2	2	3	3	4	4	5
83	9191	9196	9201	9206	9212	9217	9222	9227	9232	9238	1	1	2	2	3	3	4	4	5
84	9243	9248	9253	9258	9263	9269	9274	9279	9284	9289	1	1	2	2	3	3	4	4	5
85	9294	9299	9304	9309	9315	9320	9325	9330	9335	9340	1	1	2	2	3	3	4	4	5
86	9345	9350	9355	9360	9365	9370	9375	9380	9385	9390	1	1	2	2	3	3	4	4	5
87	9395	9400	9405	9410	9415	9420	9425	9430	9435	9440	0	1	1	2	2	3	3	4	4
88	9445	9450	9455	9460	9465	9469	9474	9479	9484	9489	0	1	1	2	2	3	3	4	4
89	9494	9499	9504	9509	9513	9518	9523	9528	9533	9538	0	1	1	2	2	3	3	4	4
90	9542	9547	9552	9557	9562	9566	9571	9576	9581	9586	0	1	1	2	2	3	3	4	4
91	9590	9595	9600	9605	9609	9614	9619	9624	9628	9633	0	1	1	2	2	3	3	4	4
92	9638	9643	9647	9652	9657	9661	9666	9671	9675	9680	0	1	1	2	2	3	3	4	4
93	9685	9689	9694	9699	9703	9708	9713	9717	9722	9727	0	1	1	2	2	3	3	4	4
94	9731	9736	9741	9745	9750	9754	9759	9763	9768	9773	0	1	1	2	2	3	3	4	4
95	9777	9782	9786	9791	9795	9800	9805	9809	9814	9818	0	1	1	2	2	3	3	4	4
96	9823	9827	9832	9836	9841	9845	9850	9854	9859	9863	0	1	1	2	2	3	3	4	4
97	9868	9872	9877	9881	9886	9890	9894	9899	9903	9908	0	1	1	2	2	3	3	4	4
98	9912	9917	9921	9926	9930	9934	9939	9943	9948	9952	0	1	1	2	2	3	3	4	4
99	9956	9961	9965	9969	9974	9978	9983	9987	9991	9996	0	1	1	2	2	3	3	4	3

TABLE Q. Monroe Table of Factors for Extracting Square Root [a]

N'	A	D	N'	A	D	N'	A	D	N'	A	D
1.000	102	202	6.29	6383	5053	19.2	1932	8791	47.35	47679	1381
1.045	1067	2066	6.49	6579	513	19.55	1978	8895	48.0	48302	139
1.09	1114	2111	6.69	6791	5212	20.0	2016	898	48.75	4914	1402
1.14	1162	2156	6.905	7017	5298	20.3	2043	904	49.5	4985	14121
1.19	1211	2201	7.132	7244	5383	20.65	2088	9139	50.35	5073	14245
1.24	126	2245	7.35	745	5459	21.0	211	9187	51.1	5148	1435
1.28	1303	2283	7.56	7681	5543	21.3	2152	9278	51.8	522	1445
1.327	1356	2329	7.80	7924	563	21.7	2186	9351	52.7	5318	14585
1.385	1416	238	8.05	8168	5716	22.1	2233	9451	53.6	54022	147
1.45	1481	2434	8.285	8404	5798	22.5	227	9529	54.5	54908	1482
1.515	1545	2486	8.53	8661	5886	22.95	2316	9625	55.2	55502	149
1.575	1609	2537	8.795	8922	5974	23.3	2343	9681	56.0	564	1502
1.645	1677	259	9.00	9102	6034	23.6	2379	9755	56.83	5738	1515
1.705	1741	2639	9.24	9345	6114	24.0	2427	9853	57.9	5843	15288
1.777	1817	2696	9.485	9619	6203	24.55	248	996	59.0	59444	1542
1.855	1896	2754	9.77	9894	6291	25.0	252	1004	59.85	6038	15541
1.94	1974	281	10.0	1009	6353	25.5	2574	10147	60.95	61465	1568
2.01	2042	2858	10.2	1032	6425	26.0	2626	10249	62.0	6252	15814
2.08	2114	2908	10.45	1053	649	26.55	26832	1036	63.1	636	1595
2.16	2177	2951	10.6	1066	653	27.1	2741	10471	64.0	644	1605
2.222	2259	3006	10.8	1086	6591	27.72	27984	1058	65.0	6548	16184
2.29	2327	3051	11.0	11122	667	28.25	2851	10679	66.0	66585	1632
2.38	2421	3112	11.2	1136	6741	28.8	29052	1078	67.2	6765	1645
2.47	252	3175	11.5	1157	6803	29.3	2955	10872	68.25	6879	16588
2.57	2621	3238	11.7	1183	6879	29.8	3003	1096	69.3	69889	1672
2.676	2729	3304	11.9	1202	6934	30.2	3047	1104	70.4	7098	1685

Value			Value			Value			Value		
2.78	2829	3364	12.1	1218	698	30.8	31138	1116	71.4	7191	1696
2.885	2936	3427	12.3	1245	7057	31.45	31753	1127	72.5	73102	171
2.99	3038	3486	12.5	1262	7105	32.05	3237	11379	73.73	7439	1725
3.09	3154	3552	12.75	1287	7175	32.7	33062	115	75.0	7549	17377
3.22	3285	3625	12.9	1305	7225	33.44	33698	1161	76.0	76562	175
3.352	3404	369	13.2	13286	729	33.9	34222	117	77.2	7788	1765
3.457	3523	3754	13.4	1352	7354	34.6	3491	11817	78.5	79032	1778
3.59	3648	382	13.7	1379	7427	35.2	35402	119	79.5	80102	179
3.71	3777	3887	13.9	1401	7486	35.78	3606	1201	80.6	8109	1801
3.80	3863	3931	14.1	1419	7534	36.3	36602	121	81.6	8221	18134
3.94	3996	3998	14.3	1436	7579	37.0	37271	1221	82.8	8321	18244
4.05	4056	4028	14.5	146	7642	37.5	37822	123	83.7	8418	1835
4.13	4188	4093	14.7	1488	7715	38.2	38502	1241	84.6	851	1845
4.26	432	4157	15.08	1528	7818	38.8	39062	125	85.5	8584	1853
4.40	4471	4229	15.45	1563	7907	39.4	3974	12608	86.4	8689	18643
4.54	4618	4298	15.8	1596	799	40.0	40322	127	87.4	8789	1875
4.705	4785	4375	16.0	1616	804	40.7	4103	12811	88.3	8883	1885
4.87	4946	4448	16.35	1652	8129	41.45	41796	1293	89.4	8968	1894
5.035	513	453	16.7	169	8222	42.1	4251	1304	90.0	9044	1902
5.225	5306	4607	17.1	1721	8297	42.9	4323	1315	91.0	9168	1915
5.39	5459	4673	17.35	1756	8381	43.7	4389	1325	92.1	9264	1925
5.56	565	4754	17.75	1788	8457	44.3	44689	1337	93.3	9406	19397
5.74	5832	483	18.05	1819	853	45.0	4536	1347	94.6	9516	1951
5.93	6017	4906	18.4	1858	8621	45.7	4609	13578	95.75	9653	1965
6.12	62	498	18.8	1897	8711	46.5	46922	137	97.1	9792	19791
6.29	6383	5053	19.2	1932	8791	47.35	47679	1381	98.6	994	1994
									100.0		

[a] For accuracy to five significant digits in roots with an error of less than 5 in the sixth digit. The use of this table is explained in Supplement A.

SOURCE: Copyright © by the Monroe Calculating Machine Company, Inc., Orange, New Jersey, 1960.

TABLE R. A Restricted Random Series of Numbers

First Hundred

Columns	Block A					Block B				
	1	2	3	4	5	6	7	8	9	10
	12	41	90	54	69	64	92	02	71	66
	57	18	53	78	30	13	00	99	28	58
	32	14	70	26	46	17	19	40	45	83
	29	82	48	42	97	91	76	88	36	37
	68	43	11	38	63	81	72	98	10	94
	85	79	16	87	96	73	34	49	47	09
	61	21	35	20	05	86	33	31	15	44
	24	06	74	80	59	22	27	51	04	89
	07	77	39	62	03	56	08	75	67	23
	93	95	01	55	84	52	60	65	50	25

Second Hundred

Columns	Block A					Block B				
	1	2	3	4	5	6	7	8	9	10
	23	19	03	68	15	59	01	88	24	40
	60	28	96	21	67	93	66	09	89	51
	49	55	80	77	11	56	38	74	34	00
	14	84	54	46	92	07	63	86	58	33
	95	29	91	72	53	90	75	52	45	82
	50	83	12	06	31	78	02	13	17	71
	47	98	79	35	04	26	97	44	61	94
	39	36	87	62	22	41	73	32	16	85
	57	64	08	48	70	20	10	18	99	27
	76	30	43	81	05	69	37	25	42	65

SOURCE: H. Friedman. A restricted random series of numbers. *Psychonomic Science*, 1966, **6**(6), 311.

TABLE S. Restricted Binary (Gellerman) Series

1	R	R	R	L	L	R	L	R	L	L		23	L	R	R	R	L	L	R	L	L	R
2	R	R	R	L	L	R	L	L	R	L		24	L	R	R	L	R	R	L	L	L	R
3	R	R	L	R	L	R	R	L	L	L		25	L	R	R	L	R	L	L	L	R	R
4	R	R	L	R	L	L	R	R	L	L		26	L	R	R	L	L	R	R	L	L	R
5	R	R	L	R	L	L	L	R	R	L		27	L	R	R	L	L	R	L	L	R	R
6	R	R	L	L	R	R	L	R	L	L		28	L	R	R	L	L	L	R	R	L	R
7	R	R	L	L	R	R	L	L	R	L		29	L	R	R	L	L	L	R	L	R	R
8	R	R	L	L	R	L	R	R	L	L		30	L	R	L	R	R	L	L	L	R	R
9	R	R	L	L	R	L	L	R	R	L		31	L	R	L	L	R	R	R	L	L	R
10	R	R	L	L	L	R	R	L	R	L		32	L	R	L	L	R	R	L	L	R	R
11	R	R	L	L	L	R	L	R	R	L		33	L	R	L	L	R	L	L	R	R	R
12	R	L	R	R	L	R	R	L	L	L		34	L	L	R	R	R	L	R	L	L	R
13	R	L	R	R	L	L	R	R	L	L		35	L	L	R	R	R	L	L	R	L	R
14	R	L	R	R	L	L	L	R	R	L		36	L	L	R	R	L	R	R	L	L	R
15	R	L	R	L	L	R	R	R	L	L		37	L	L	R	R	L	R	L	L	R	R
16	R	L	L	R	R	R	L	R	L	L		38	L	L	R	R	L	L	R	R	L	R
17	R	L	L	R	R	R	L	L	R	L		39	L	L	R	R	L	L	R	L	R	R
18	R	L	L	R	R	L	R	R	L	L		40	L	L	R	L	R	R	R	L	L	R
19	R	L	L	R	R	L	L	R	R	L		41	L	L	R	L	R	R	L	L	R	R
20	R	L	L	R	L	R	R	R	L	L		42	L	L	R	L	R	L	L	R	R	R
21	R	L	L	R	L	L	R	R	R	L		43	L	L	L	R	R	L	R	R	L	R
22	R	L	L	L	R	R	L	R	R	L		44	L	L	L	R	R	L	R	L	R	R

SOURCE: L. W. Gellerman. Chance orders of alternating stimuli in visual discrimination experiments. *Journal of Genetic Psychology,* 1933, **42**, 207–208.

references

COHEN, J. Some statistical issues in psychological research. In B. B. Wolman (Ed.), *Handbook of Clinical Psychology*. New York: McGraw-Hill, 1966.

DUKES, W. F. $N = 1$. *Psychological Bulletin*, 1965, **64**(1), 74–79.

EDGINGTON, E. S. Statistical inference from $N = 1$ experiments. *The Journal of Psychology*, 1967, **65**, 195–199.

FLEISS, J. L. Estimating the magnitude of experimental effects. *Psychological Bulletin*, 1969, **72**(4), 273–276.

FRIEDMAN, H. Colour vision in the Virginia opossum. *Nature*, 1967, **213**(5078), pp. 835–836.

FRIEDMAN, H. Magnitude of experimental effect and a table for its rapid estimation. *Psychological Bulletin*, 1968, **70**(4), 245–251.

FRIEDMAN, H. A restricted random series of numbers. *Psychonomic Science*, 1966, **6**(6), 311.

FRIEDMAN, H. A simplified table for the estimation of magnitude of experimental effect. *Psychonomic Science*, 1969, **14**(4), 193, 195.

FRIEDMAN, M. The use of ranks to avoid the assumption of normality implicit in the analysis of variance. *Journal of the American Statistical Association*, 1937, **32**, 675–701.

GELLERMAN, L. W. Chance orders of alternating stimuli in visual discrimination experiments. *Journal of Genetic Psychology*, 1933, **42**, 206–208.

GRIZZLE, J. E. Continuity correction in the χ^2-test for 2×2 tables. *The American Statistician*, October, 1967, 28–32.

GUILFORD, J. P. *Psychometric Methods*. New York: McGraw-Hill, 1954.

HAYS, W. L. *Statistics for Psychologists*. New York: Holt, Rinehart and Winston, 1963.

KIRK, R. E. *Experimental Design Procedures for the Behavioral Sciences*. Belmont, Calif.: Brooks-Cole Publishing Co., 1968.

KRUSKAL, W. H., and WALLIS, W. A. Use of ranks in one-criterion variance analysis. *Journal of the American Statistical Association*, 1952, **47**, 614–617.

LEVY, P. Substantive significance of significant differences between two groups. *Psychological Bulletin*, 1967, **67**, 37–40.

MANN, H. B., and WHITNEY, D. R. On a test of whether one of two random variables is stochastically larger than the other. *Annals of Mathematical Statistics*, 1947, **18**, 50–60.

RYAN, T. A. The experiment as the unit for computing rates of error. *Psychological Bulletin*, 1962, **59**, 301–305.

SIEGEL, S. *Nonparametric Statistics for the Behavioral Sciences*. New York: McGraw-Hill, 1956.

STEVENS, S. S. On the theory of scales of measurement. *Science*, 1946, **103**, 677–680.

WHITE, C. The use of ranks in a test of significance for comparing two treatments. *Biometrics*, 1952, **8**, 33–41.

WILCOXON, F. *Some Rapid Approximate Statistical Procedures*. Stamford, Conn.: American Cyanamid Co., 1949.

WINER, B. J. *Statistical Principles in Experimental Design.* New York: McGraw-
Hill, 1962.

WOOD, D. A. *Test Construction.* Columbus, Ohio: Charles E. Merrill Books, 1961.

TYPE OF STUDY

Type of Data	Single Sample	Two Samples	
		Independent Groups	Related
			Differences
Category Data	Binomial A. p. 6 B. p. 10 Table H, p. 290	χ^2 Multi-sample A. p. 29 B. p. 42 Table F, p. 288	* (Mc Nemar Test)
	χ^2 Single Sample A. p. 15 B. p. 24 Table F, p. 288	* (Fisher Exact Test)	
Ranked Data	χ^2 Single Sample A. p. 15 B. p. 24 Table F, p. 288	White A. p. 51 B. p. 62 Table I, p. 292	Matched-Pairs Signed-Ranks A. p. 57 B. p. 64 Table J, p. 295
	* (Kolmogarov Smirnov)		
Score Data	t Single Sample A. p. 132 B. p. 148 Table A, p. 269	t Independent Sample A. p. 135 B. p. 150 Table A, p. 269	t Related Sample A. p. 141 B. p. 152 Table A, p. 269

A. Text Coverage
B. Worked Example

Additional Tables

Table B r_m Simplified, p. 270
Table C r_m Expanded, p. 273
Table D Normal Distribution, p. 280
Table G ϕ Coefficient, p. 289